SP Pvt. Ltd.

Mastering
Life Science

by
Midhila Padman
Department of Biotechnology
Sahrdaya College of Engineering and Technology
Thrissur, Kerala – 680 684

2020

Studium Press (India) Pvt. Ltd.

Mastering

Life Science

ISBN: 978-93-85046-67-4

Published by:

Studium Press (India) Pvt. Ltd.
4735/22, 2nd Floor, Prakash Deep Building
(Near Delhi Medical Association)
Ansari Road, Darya Ganj, New Delhi-110 002
Tel.: + 91-11-43240200-15 (15 lines); Fax: 91-11-43240215
E-mail: pubdir@studiumpress.in

Printed at India

ABOUT THE AUTHOR

Dr. Midhila Padman holds a PhD from Mysore University. She has a vast and variegated experience of teaching and research in the field of Life Science. Dr. Midhila is currently working as Assistant Professor in Biotechnology Engineering. Her Research interests lie in Plant Pathology, Environmental Biotechnology and Rural development. She has worked with DBT Citizen Fellowship in the field of Environmental Health and has qualified NET. She has also contributed research papers in reputed journals of National and International standards and has authored book in Malayalam Literature. She is also a constant investigator on Life Science.

PREFACE

Human life has enormously been influenced by advances in Science and Technology. Life is the prime concern of every species, but understanding the concepts of life can at times be ambiguous and may or may not define it in proper perspective of the learner. Life science is concerned mainly with understanding life processes in terms of science and this book is indented for students and learners who wants to learn about the science of life and technological advancements in science with a most exciting and straightforward presentation of Biotechnology, Microbiology, Genetics, Genetic Engineering, Industrial Biotechnology and other recent approaches in Life science. Currently there is an increased interest in the study of these subjects because investigations based on cells including microbial, human and animal cells provide new lead for human development and sustainability of environment. Technologies based on Biotechnological and Genetical approaches find increasing application in day today life. Hence understanding the concepts in Life science to compete with the multiple academic disciplines and to fulfil the aspirations of learners for the benefit of society and environment can be achieved through inheritance of the subject knowledge. This book would cater to the need of graduate, post graduate students pursuing their study in all branches of life sciences and also for the students who are in a process of facing competitive examinations like NET etc.

INTRODUCTION TO LIFE SCIENCE

IMPORTANCE OF LIFE SCIENCE

The **life sciences** or **biological sciences** comprise the branches of science that involve the scientific study of life and organisms such as microorganisms, plants, and animals including human beings. Life sciences discoveries are helpful in improving the quality and standard of life, and have applications in health, agriculture, medicine, and the pharmaceutical and food science industries.

Genetics is the study of inheritance of characters or heredity and variations. Heredity is the study of expression and transmission of traits from parents to offspring. The application of Genetics is significant as any cell based exploitation or innovative strategies in the field of medicine, healthcare or environment requires a wise study of genetics. A closely related branch **Genetic engineering** is defined as the direct manipulation of an organism's genes including heritable and nonheritable recombinant DNA constructs.

Environmental biotechnology is the branch of biotechnology that addresses environmental problems, such as the removal of pollution, renewable energy generation or biomass production, by exploiting biological processes and hence is the recently utilized branch of Biotechnology.

Industrial Biotechnology and fermentation technology or white **biotechnology** uses microorganisms and enzymes to produce goods for industry, including chemicals, plastics, food, agricultural and pharmaceutical products and energy carriers and is highly significant and widely accepted branch of Science.

TABLE OF CONTENTS

About the Author v
Preface vii
Introduction to Life Science viii

1. Genetics **1**

1.1. Organization of Prokaryotic DNA 1
1.2. Organization of Eukaryotic Chromosome 2
1.3. Cell Division 5
1.4. Bacterial Recombination 13
1.5. Identification of DNA as Genetic Material 24
1.6. Types of RNA 29
1.7. Types of DNA 33
1.8. DNA Replication 34
1.9. Transcription 38
1.10. Translation 41
1.11. Genetic Code 44
1.12. Regulation of Gene Expression 46
1.13. Mutations and Repair 49
1.14. Gene Protein Relationship 56

2. Enzymology **58**

2.1. Enzymes 58
2.2. Factors affecting Enzyme Activity 59

3. Summary of Important Pathways **67**

3.1. Electron Transport and Oxidative Phosphorylation 67
3.2. TCA Cycle (Citric Acid Cycle) 70
3.3. Biosynthesis of Fatty Acid 74

4. Genetic Engineering **79**

4.1. Cloning Host 79
4.2. Transformation of *E Coli* 96

4.3. Electrophoresis 99
4.4. Blotting Techniques 102
4.5. Methods of DNA Sequencing 106
4.6. Introduction of Recombinant DNA into Host Cells 108
4.7. Selection of Recombinant Bacteria 111
4.8. Important Molecular Techniques 113

5. Environmental Biotechnology 116

5.1. Biological Indicators 116
5.2. Water Borne Diseases 119
5.3. Determination of Sanitary Quality of Water 124
5.4. Municipal Water Treatment 128
5.5. Sewage Microbiology 135
5.6. Waste Water Treatment 144
5.7. Municipal Waste Water Treatment 147
5.8. Anaerobic Digestion 161
5.9. Reclamation of Waste Water 168

6. Industrial Biotechnology 171

6.1. Strain Improvement 171
6.2. Production Medium 175
6.3. Raw Materials for Process Control 180
6.4. Design of Fermentor 183
6.5. Types of Fermentation 187
6.6. Down Stream Processing 191

7. Electron Microscopy 199

8. Control of Micro Organisms 202

9. Mycotoxins 207

Suggested Reading 211

Subject Index 213

Chapter 1

GENETICS

1.1. ORGANIZATION OF PROKARYOTIC DNA

Prokaryotic cells lack membrane bound nucleus and have no distinct nuclear compartment. The Prokaryotes are divided into eubacteria or true bacteria including *E. coli* and unusual group including Archaebacteria.

Prokaryote genomes are exemplified by *E. coli* chromosome. The bulk of DNA in *E. coli* cells consists of a single closed circular DNA molecule of length 4.6 million base pairs. The DNA is packaged into a region of cell known as nucleoid. This region has a very high DNA concentration as well as containing all proteins associated with DNA such as Polymerases, repressors and others (In normal growth DNA is being replicated continuous and there may be on average around two copies of genome per cell, when growth is at maximal rate).

DNA consists of 50–100 domains or loops, the ends of which are constrained by binding to a structure which consist of proteins attached to part of cell membrane. Loops are 50–100 kb in size. The chromosome as a whole is negatively supercoiled. The attachments of DNA to protein membrane scaffold may act as a barrier to rotation of DNA, such that the domains may be topologically independent.

The looped domains of DNA are constrained further by interaction with a number of DNA binding proteins. The most abundant of these are protein HU, a small basic (positively charged) dimeric protein, which binds DNA nonspecifically by wrapping of the DNA around protein and

H-NS, a monomeric neutral protein which also binds DNA nonspecifically. This proteins are sometime known as histone like proteins and have an effect of compacting DNA, which is essential for packaging DNA into nucleoid and stabilizing and constraining the supercoiling of chromosome.

RNA polymerase and mRNA molecule as well as site specific DNA binding proteins such as Integration Host Factor (IHF), a homolog of HU (which bind to specific DNA sequences and binds DNA) may be important in the organization of DNA domains.

The DNA molecule is extremely long related to dimensions of cell. To allow it to fit inside cell, it is compacted by a process called *Supercoiling*. Enzyme Topoisomerases introduce additional turns into double helix that causes DNA strands to wind upon itself and adopt more compact form. The Topoisomerases a act by braking DNA polynucleotide and rotating the two ends relative to each other. The enzyme then the rejoins the ends of polynucleotide reacts by winding upon itself. This is called *positive supercoiling*. Topoisomereases can also remove coiling through process called *negative supercoiling* by creating a turn in opposite direction. It has a central protein core from which supercoiled DNA loops radiate.

Prokaryotic genes

Almost all genes present in bacteria occur on bacterial chromosome. A few other genes rest on small circular DNA molecules called *plasmids*. *E. coli* have 4300 genes carried by 4.6 million base pairs. Some genes are arranged as families and are regulated on a coordinated way, but most of the genes are present in single copies. The main exception is in ribosomal RNA genes that occur in cluster of seven copies.

About 90% of DNA of bacterial chromosome is accounted for by genes. The remaining 10% is intergenic DNA which separates individual genes. Some parts of intergenic DNA have important functions such as location of Origin of replication of bacterial chromosome. Other intergenic regions may be involved in interactions with DNA packaging proteins

1.2. ORGANIZATION OF EUKARYOTIC CHROMOSOME

Analysis of chromosome of eukaryotic cells has shown them to be composed of DNA and protein with small amounts of chromosomal RNA.

DNA has negative charges distributed along its length and positively charged (basic) proteins molecules called Histones bind to it. This DNA protein complex is called chromatin.

Eukaryotic cells contain more than one set of chromosomes each set is composed of several different linear chromosomes. A human cell for example contains 2.2 m of DNA distracted among 46 chromosomes. Each chromosome is on average about 6 μm long. In order to maintain the high degree of organization when DNA is folded, the Histone protein form precise architectural scaffolding for DNA.

Chromatin Structure

The DNA helix combines with group of protein molecules called *Histone protein* to form Nucleosomes. The repeating structural unit within Eukaryotic chromatin is the *Nucleosome*. As proposed by Roger Kornberg (1974), the Nucleosome is composed of double stranded DNA wrapped around an Octamer of Histone proteins.

The chromatin of Eukaryotic cells contain a repeating patterns in which the Nucleosomes are connected by linker regions of DNA that vary in length from 20–100 bp. The DNA lies on the surface and takes 1–8 negative supercoils (helical turns) around the Histone octamer. The amount of DNA that is required to wrap around the histone octamer is 145 base pairs. The entire chromosome is 11 mm in diameter.

In Eukaryote chromatin Nucleosomes are associated with each to form an even more compact structure, 30 nm in diameter. The 30 nm fiber shortens the total length of DNA another seven fold. This 30 mm fiber is also known as *solenoid fiber*.

A third mechanism that compact the DNA further involves the formation of radial loop domains. In this process chromatin is organized into loops often 50,000–100,000 p in size that are enclosed to a nuclear scaffold, known as nuclear matrix. The nuclear scaffold is composed of many cellular proteins that bind to each other and to DNA. These scaffolding proteins come together to form fairly rigid structure that provides a support for attachment of loop domains.

The DNA sequence that bind to the nuclear scaffold are known as *Scaffold Attached Regions (SARs)* or *Matrix Attachment Regions (MARs)*. These regions are few hundred nucleotides in length and hence have a high percentage of A and T residues.

Role of Histones in Nucleosome Organization

Histone proteins are very basic proteins because they contain a large number of positively charged lysine and arginine residues. The arginic residues play a major role in binding to DNA. Arginic residues within histone proteins form electrostatic and hydrogen bonding interactions within the phoshate groups along DNA backbone.

There are four families of core histones- H2A, H2B, H3, H4 and a further family H1. A single molecule of H1 stabilizes the DNA at the point at which it enters and leaves the necleosome core and organizes the DNA between nucleosomes. A nucleosome core plus H1 is known as Chromatosome. In some cases H1 is replaced by a varient-H5 which binds more tightly and is associate with DNA which is inactive in transcription.

The histone core consist of the octomer of 8 histone molecules organized in the form of two flat tetramers of H2A, H2B, H3 and H4.

The genome of eukaryotes contain sequences that are unique, moderately repetative and repetative. The term sequence complexity refers to the number of times a particular base sequence appear through out the genome.

Unique or non repetative sequence are those found only once or a few times within the genome. The vast majority of proteins in eukaryotic cells are encoded by genes that are present in one or few copies. In case of humans 60–70 of entire genome constitutes unique sequencc.

Moderately repetative sequences are those found a few hundred to several thousand times in genome. *Eg*: Gense that encode ribosomal RNA (rRNA)

Highly repetative sequence are found in tens of thousand or ever millions of times throug out the genome. *Eg*: Alu sequence (300 bp in length approximately) found in humans.

Some highly repetative sequences are clustered together in tandem array. In tandom array a very short nucleotide sequence is repeated many times in a row. Tandem arrays of highly repetative sequences are found in centromeric regions of chromosomes.

1.3. CELL DIVISION

In prokaryotes such as bacteria, partitioning of the nuclear material (*i.e.*, Bacterial chromosome) occur by utilizing the attachment of the chromosome and its newly formed replicate to a section of bacterial cell membrane. As the bacteriemn elongates duing cell devision, one portion of the membrence carrying one chromosome seperates from another portion carrying its replicate. When sufficient seperation between the two chromosomes have been achieved, the membrence between them invaginates and cleaves, so that each of the chromosome is now in seperate daughter cell. In eukaryotie cells, division is more complex because more than one chromosome is present.

Mitosis

Interphase

The interphase period between sucessive cell devision consist of processes associated with growth and prepartion for mitosis. In many case the specific chemicals that consititute the newly sythesized chromosome (DNA and histone) as well as the proteins that will soon give rise to the mitotic spindle are found during this period. The period of DNA synthesis during interphase is called S (synthetic) period and is seperated in time from the previous cell division by a gap called G1. After DNA synthesize, a further gap called G2 occur before the next cell division begins.

The G1 period shows considerable variability often raging from 3–4 hours to week or months. The G2 shows more cosistancy and ranges from 2–5 hours and S is consistant, which lasts about 7–8 hours.

Prophase

In prophase the first stage of mitosis, coiling of chromosome or their condensation begins making them visible as thread like structure. Each of the mitotic prophase chromosome appears longitudinally split into two duplicates each of which is called a chromatid as long as it remains connected to its sister chromatid. Further prophase event are splitting of centrosome and movement of each of each half (centriole) to opposite side of the nucleus, the synthesis of mitotic apparatus, the disappearance of nucleus and beginning of breakdown of nuclear membrane. Cell cycle showing S, M and G1 intervals (gap between cell divisio) G1 is variable, whereas G2 and are fairly constant.

Metaphase

Once the new chromosome material has been synthesized and the chromosomes have coiled and condensed, they began succession of active movements accompanied by complete breakdown of nuclear membrane. These movements are based on the attachment of each chromosome though a specific point along its length, the centromere (also called *kinetochore*) to a double poled spindle shaped structure the spindle. The centromere is characterized by localized construction that occupies a constant position for any chromosome, but some special process diffuse centromeres with numerous spindle attachment points. Chromosome whose centromere is localized approximately midway between each end, thereby forming two equal chromosome "arms" are described as metacentric chromosomes, while the one with a more terminally placed centromere forming unequal chromosome "arms" called acrocentric. Some chromosome material always exists on both sides of centromeres, where some centromere appears to be at the very tip of chromosome, in such case it is called *telocentric*. Metaphase configuration showing each chromosome split into two sister chromatids and remains connected at centromere region.

Anaphase

The shortest of the mitotic phase, anaphase occur when the mutual attachment between the two sister chromatids ceases. At that point, the centromere of each chromatid on the metaphase plate seperates from its sister centromere and the products of this division, which may now be called daughter chromosome move toward opposite poles of the spindle. The appearance of this event is that of an active repulsion between former sister centromeres, each dragging its chromosome along with it.

Telophase

During telophase period each of the two polar groups of daughter chromosome undergoes a reversion to more extended interphase state. The nuclear membrane is re-established and nucleoli formed. Division of the cytoplasm protein of the cell (cytokinesis) is also completed during this period giving rise to a cell plate in plant cells and indented cell furrow in animal cells

Meiosis and Homologous Pairing

In sexually reproducing organisms where zygote is formed by fertilization between male and female sex cells or gametes, the embryonic cells would double the chromosomes of each part. Sexually reproducing organs has evolved a mechanism that enables them to regularly reduce the member of chromosome in each gamete to half the sensual member.

Meiosis is a process by which the chromosome is seperated during the formation of sex cells and their numbers reduced from diploid to haploid condition.

First Meiotic Prophase

Meiotic division follows two recessive divisions of chromosomes. The first division represents a reduction division in which membrane of homologous pairs of chromosomes are separated into daughter cells without duplication *i.e.*, their numbers are reduced to half. During the fairly lengthy first meiotic prophase, the events have been divided into five stages.

Laptotene

The Chromosome first appear as long slender beads with many bead like structure (chromomeres) along their length. Meiotic prophase chromosome generally appear as single and individual structure (without any morphological duplication) until the pachytene stage.

Zygotene

During zygotene stage homologues chromosomes appear to attract each other and enter into a very close zipper like pairing (*synapsis*) This pairing is highly specific and occurs between all homologues chromosome sections even if present on different non homologues chromosome.

Pachytene

The panchytene is a stage of progressive shortening and coiling of chromosome that occur once zyogotic paring has been completed. At pachytene two sister chromatids of homologues chromosome are associated with two sister chromatids of their homologues partner. This group of four chromatids is known as bivalent or tetrad and a series of exchanges of genetic material can occur between nonsister homologues chematids. Such exchanges results in crossing over and recombination.

Pachytene is also the stage in which a structure called synaptonemal complex can be observed between synapse chromosomes through electron microscopy. It appears as ribbon like group of three longitudinal components organized in two dense lateral elements and a thin central element composed primarily of proteins. The synaptonemal complex may function to pull chromosomes together helping them to pair more precisely and efficiently. Under light microscope, an easily observable counterpart to genetic crossing over, the chiasmata can be noted.

Diplotene

At the diplotene stage each chromosome acts as though it were repulsing its closely paired homologue near centromere. Distinctly visible separations then occur between homologues chromosomes except for specific regions, where an actual physical crossing over appears to have taken place between homologous chromatids. These crossed areas or chiasmata are X-shaped attachments between chromosomes and seem to the only remaining force holding each bivalent together until metaphase.

Diakinesis

At diakinesis coiling and contraction of the chromosomes continue until they are thick heavy staining bodies. In this process the bivalents usually migrate close to thc nuclear membrane and become evenly distributed. The nucleolus either disappears or detaches frem its associated chromosome during the latter part of this stages or early part of metaphase. The nuclear membrane dissolves and the bivalents attach themselves by their centrioles to rapidly formed spindle.

First metaphase

Terminalization of chiasmata in bivalent show reduction of the chiasmata in diplotene to terminal chiasmata. In metaphase the chromosomes reach their most condensed state and appear relatively smooth in outline. The chiasmata that had first appeared during diplotene have now moved towards the ends of each chromosome (terminalization) leaving only the single terminal attachment between the formerly paired arms of homologues chromosome. These remaining chiasmata prevent separation of homologous chromosome which now lie on each side of equatorial plate of spindle stretched by their respective centromere towards opposite poles.

First Anaphase

Athough previously duplicated along its entire length, each chromosome still maintains single functional centromere for both of its sister chromatids. The separation or disjunction of or homologues chromosome from another in anaphase towards opposite poles therefore results in single chromosome centromere, which may already be structurally split dragging both chromatid (dyad) along with it, forming two haploid sets. Each chromosome still contains a pair of sister chromatids.

First Telophase

The last meiotic division effectively ends when the chromosomes arrive at the poles. Each daughter cell now has half the numbers of chromosomes, but each microtubule that make up the spindle network disappear and new nuclear membrane surrounds each haploid set. The chromosome uncoils back into chromatin. Sister chromatids remain attached during telophase I

Cells may enter a period of rest known as interkinesis or interphase II. No DNA replication occurs during this stage. Meiosiss II is second part of meiotic process. Much of the process is similar to mitosis. The four mains steps of meiosis II are

1. ***Prophase II*:** Disappearance of nucleoli and nuclear envelop again as well as shortening and thickening of chromatids. Centrioles move to polar regions and arrange sprindle fibers for second meiotic devision.
2. ***Metaphase II*:** Two dyad on spindle in each daughter cell
3. ***Anaphase II*:** Where centromere is cleaved and sister chromatids pull apart (the sister chromatids by convention are now called sister chromosome as they move towards opposite poles.
4. ***Telophase II*:** Marked by uncoiling and lengthening of chromosomes and disappearance of spindle. Nuclear envelop reform and cleavage or cell wall formation eventually produces a total four daughter cells, each with haploid set of chromosomes.

Genetics was a young science branch, since its founding father Gregor Mendel performed his ground breaking work. Mendals work revealed two fundamental truths: that physical traits are determined by factors (now called genes) passed on by both parents from one generation to other. Mendal's contributions are so profound that his ideas are often referred to as Medel's laws of inheritance

Mendel tested seven characters individually by crossing a variety carrying a particular trait or character *Eg:* seed color and seed shape, tall and short. When he crossed smooth seed shape variety with wrinkled once he obtained seeds that were all smooth; similarly when he crossed yellow seed plant to green seed variety, seeds produced were all of type yellow. An initial cross between two varieties is called Parental or P1 generation and their off spring; called F_1 generation. Succeeding generations descended from their cross are labeled F_2 and so on.

Mendel's crosses between two different varieties for each character always produced on F_1 that was only one type. However when these F_1 plants were permitted to reproduce by self fertilization, examples of both original varieties now appeared in F_2.

These F_2 ratios are very close to 3:1 and therefore can be written 3/4:1/4 or 75:25 or 75 percent:25 percent.

The result for all 7 characters appeared in the following pattern

1. For any character F_1 derived from crosses between two different varieties showed only one of the traits, never the other.
2. The trait that has disappeared or been hidden in F_1 reappeared in F_2 in a frequency one quarter that of total number.
3. It did not matter which parent variety produced the pollen and which the ova: reciprocal crosses in which each of two varieties used to provide male and female parents always gave the same result. These three was no indication that any trait Mendel observed was transmitted only through meternal or paternal inheritance, but rather any trait would be passed by both parents.

The phenomenon by which one trait appears and other does not, even though both factors are present is called *dominance* (*Eg:* smooth dominant over 'S' Wrinkled) and other one as *recessive* ('s').

When smooth offsprings produced by F_2 generation were self fertilized, they intern give rise to same types of plants- One third pure smooth producers and two third hybrid smooth wrinkled producers. The types of plants could be distinguished in these crosses as pure smooth producers, pure wrinkled producers and wrinkled smooth producers.

The random combination of gametes to form Zygotes, observed in ratio showed hybrid between two different varieties posess both types of parental factors which subsequently separate or segregate in gametes. This fundamental law is known as *principle of segregation*.

Terminology

Allele: One alternative form of a given allelic pair or the two individual genes in a particular gene pair. Tall and dwarf are the alleles for height in pea plant.

Allelic pair: The combination of two alleles which comprise the gene pair

Homozygote: An individual which contains only one allele at the allele pair *Eg:* DD is homozygous dominant and dd homozygous recessive.

Heterozygote: An individual which contain one of each member of the gene pair *Eg:* Dd

Genotype: The specific allele combination for certain gene or set of genes.

P_1	Smooth X Wrinkled	Yellow X Green
F_1	Smooth	Yellow
F_2	¾ Smooth:1/4 Wrinkled	¾ yellow:1/4 Green

Here each gene pair acts independently of other. This means that the chance for a plant to be smooth or wrinkled do not interfere with or are independent of its chance to be yellow or green. Mathematically this relationship is expressed by multiplying the probability that plant will have a certain characteristic *Eg:* Smooth, times the probability that a plant will have certain charecteristic *Eg:* Yellow if the seed has ¾ chance of being smooth and ¾ chance of being yellow its chance to be smooth or yellow at the same time would be $3/4 \times 3/4 = 9/16$ we can therefore obtain the ratio of each phenotypic combinations by multiplying the expected ratio by total number. Since these are almost exact ratios and numbers observed, they testify the independence or independant assortment of two gene pair.

Since smooth is dominant over wrinkled and yellow over green F_2 plants that arose from wrinkled and green seeds must be homozygous for both recessive and have genetic constitution ssyy (double recessive).

Such plants upon self fertilization would be expected to produce only wrinkled green seeds (exactly what Mendel found). Smooth and yellow plants on the other hand might have varied genotypes SSYY, SsYY, SSYy, SsYy

SsYy	X	ssyy	(Wrinkled green parent)
SsYy	X	SSYY	(Smooth yellow parent)

Although F_1 is expected to produce four types of gametes in equal ratio (SY, Sy, sY and sy) the double receive parental wrinkled green strain in cross 1 will produce only sy gametes.

Gametes of Wrinkled green variety

F_1 Gametes

SY	Sy	sY	sy
SsYy	Ssyy	ssYy	ssyy

Thus four phenotypic combinations of offspring would be expected in equal proportion. The four phenotypic classes appeared in a 1:1:1:1 ratio.

The second cross between F_1 hybrid and smooth yellow variety produced off spring of only one phenotype smooth yellow.

Gametes of smooth yellow variety

F_1 Gametes

SY	Sy	sY	sy
SSYY	SSYy	SsYY	SsYy

Segregation and Assortment in Haploid Organisms

The major portion of life cycle in organisms such as algae and fungi is in haploid stage. Such organism becomes diploid during sexual conjugation, but then only until meiosis forms new haploid clones. Since the presence of chromosomes in haploid condition is equivalent to presence of genes in unpaired condition, haploid organisms are neither homozygotes nor heterozygotes. Phenotypically therefore each allele in haploid appears undistinguished by an allele mate and segregation and assortment between genes can be expected to give different phenotypic ratios than in diploid organisms *Eg:* Single celled

1.4. BACTERIAL RECOMBINATION

Genetic recombination is the formation of a new genotype by reassortment of genes following an exchange of genetic material between two different chromosomes which have similar genes at corresponding sites. These are called homologous chromosomes and are from different individuals. Progeny from recombination have combinations of genes different from those that are present in parent.

In bacteria, genetic recombination results from three types of gene transfer.

1. ***Conjugation*:** Transfer of genes between cells that are in physical contact with one another.
2. ***Transduction*:** Transfer of genes from one cell to another by a bacteriophage.
3. ***Transformation*:** Transfer of cell free or naked DNA from one cell to another.

These three types of genes transfer do not have fusion of cells, and usually only a portion of chromosomes from donor cell (male) is transferred to the recipient cell (female). The recipient cell therefore become a merozygote; a zygote that is partial diploid. Inside the recipient DNA in such a way that homologues genes are adjacent. Enzymes act on the recipient DNA, causing nicks and excision of a fragment. Then donor DNA is integrated into recipient chromosome in place of excised DNA. The recipient cell then becomes recombinant cell because its chromosomes contains DNA of both donor and recipient cell (the excised DNA pieces from recipient chromosomes are broken).

Bacterial Conjugation

The first demonstration of recombination was achieved by Leaderberg and Tatum in 1946. To select a few possible recombinations out of large population, they combined two different auxotrophic strains of *E. coli* and gave them an opportunity to male. Then they plated combined cultures on a minimal medium, where only prototrophs could grow. When they found prototrophic colonies growing there it could asserted that recombination between the auxotroph has taken place.

NOTE: Lederberg and Tatum used polyauxotroph with more than one nutritional requirement so that back mutation or spontaneous mutation to the wild type would not occur.

It had been shown that the prototrophs arose not because of transformation. Cell to cell contact as a precondition was shown by using U tube, where the auxotrophic parental cells could be cultured together yet separated physically (by microporous fritted glass permissible to molecules but not cells). Macromolecules like DNA are allowed to pass freely but not cells. Also culture fluid and soluble material could pass freely between two parent cultures. Prototroph or recombinants were never recorded from such cultures. These result indicated that bacterial conjugation is indeed a true sexual process.

While only very small fragments of bacterial chromosome are transferred in transduction and transformation, in conjugation it is possible for larger segments of chromosome (or in special cases the entire chromosome) can be transferred.

Sex factors

There exist different mating types of bacteria. Male cells contain a small circular piece of DNA, which is in the cytoplasm and not a part of chromosome called the sex factor or F factor (fertility factor). These cells are referred to as F+ and act as donors in mating. Female cells lack this factor and are labeled F–. They act as recipient cells.

Crosses between two F– strains do not yield recombination. However in F+ × F– crosses, male replicates its sex factor and one copy of it is almost always transferred to female receipt. The F– cell will be converted to F+ cells. The transfer of F factor in an F+ × F cross occur with a frequency that approach 100% but the formation of recombination in F+ × F cross occur at low frequency about one recombinant per 10^4 to 10^5 cells. One or more sex pili are produced by each F+ cell. Sex pilus act to bind F– cells to an F+ cell.

$F^+ \times F^-$ mating

During $F^+ \times F^-$ mating F factor replicates by rolling circle mechanism and a copy moves to recipient. The entering stand is copied to produce double stranded DNA. Because bacterial chromosome genes are rarely transferred with independent F factor, recombination frequency is low. It is still not completely clear how plasmid moves between bacteria. The sex pilus joins donor and recipient and may contract to draw them together.

Eg: Bacillus, Streptococcus, Enterococcus, Staphylococcus.

Hfr Conjugation

Because certain donor strain transfer bacterial genes with great efficiency and do not usually change recipient bacteria to donors, a second type of conjunction exists. The F factor is an Episome and can integrate into bacterial chromosome at several different locations by recombination between homologues sequence present on both plasmid and host chromosomes.

Plasmid can direct synthesis of pili, carry out rolling circle replication and transfer genetic material to F- recipient. Such a donor is called Hfr strain (Hfr stands for high frequency of recombination) because it exhibits a high frequency of chromosomal gene transfer in comparison with F+ cells. DNA transfer begins when the integrated F factor is nicked at its site of transfer origin. As it is replicated, chromosome moves through pilus to recipient from donor. Because only a part of F factor is transferred at the start, F- receipent does not become F+ unless whole chromosome is transferred. Transfer is standardized at 100 minutes in *E. coli* and the connection usually breaks before this process is finished. Thus a complete F factor usually is not transferred and recipient remains F-.

After replicated donor chromosome enters recipient cell. It may be degraded or incorporated into F- genome by recombination

F' Conjugation

Because F plasmid is an episome, it can leave the bacterial chromosome. Sometimes during this process, plasmid makes an error in excision and pickup a portion of chromosome material to form F' plasmid. An inclusion of one or more genes in excised F plasmid takes place and F' cells retains all of its genes although some of them are on plasmid and still mates only with an F– recipient. $F^1 \times F-$ conjunction is virtually identical with $F+ \times F-$ mating. The receptor become F_1 and is partially merozygote, since it has two sets of genes carried by plasmid. In this way specific bacterial genes may spread rapidly throughout bacterial populations. Such transfer of bacterial genes is often called *sexduction*.

Transduction

Most bacteriophages; the virulent phages, undergo a rapid lytic growth cycle in their host cells. They inject their nuclear material (usually DNA) into bacterium where it replicates rapidly and also directs the synthesis of new phage proteins. Within 10–20 (depending on phage) minutes, new DNA combines with new proteins make whole phage particles, which are released by destruction of cell wall and lysis of cell. However some bacterial viruses, the temperate phages, which ordinarily do not lyse the cell, carry DNA that can behave as episome in bacteria. Like other episome such as F factor, these viral genomes can become integrated into bacterial genome- they are known as prophage. Bacteria that carry prophage (lysogenic bacteria) can be induced with UV light or other agents to make the prophage start to replicate rapidly and go through lytic growth cycle, resulting in lysis and release of new phage particles. Such aberrant phages can attach to other bacteria and introduce bacterial, rather than just phage DNA into them; by this means they transfer bacterial DNA from one cell to other. Bacteriaphage thus serve as a vector to transfer DNA from one bacterium (donor) to another (recipient).

This phenomenon was discovered by Zinder and Lederberg in 1952 when they searched for sexual conjugation among *Salmonella* species. They mixed auxotroph mutant together and isolated prototrophic recombinant colonies from selective nutritional media. When U tube experiments were carried out with parental auxotrophs in each arm separated by microporous fritted glass filter prototrophs appeared in one arm of tube. Since the filter prevent cell to cell contact but allowed free passage of fluid between cultures, it could be concluded that some phenomenon other than conjugation was involved. This phenomenon could not be stopped with DNase activity, thus eliminating transformation as the process for changing the recipient auxotrophs to prototrophs. Further experiments implicated bacteriophage as vector or transuding agent for the transfer of gentic material – The phage passed though filter and infected donor cells lysing it. During replication in donor strain phage included parts of chromosome with it. It then passed through filter carrying a part of genetic information and imparting it to recipient strain.

Generalized Transduction

In generalized transduction random portion of bacterial chromosome are transferred from donor to recipient cell. Generalized transduction can occur in any population of bacterial cells susceptible to infection by bactreriophage. The events could be summarized as,

- Following infection of host bacteriophages often initiate lytic cycle in the phage. Nucleic acid is replicated and packaged into phage particles. Under normal conditions these phage particles are released upon lysis of host cell.

- Occasionally, host chromosome can be fragmented into small portions (fragments that contain only 1–2% of total bacterial DNA) during the lytic cycle and DNA fragmenst are mistakenly packed into some of progency phage particles which are released upon cell lysis.
- The lysate produced by generalized transduction contain mixture of phages containing random fragments of bacterial chromosome and phages with only phage nucliec acid.
- Transducing phage particles containing host DNA have very little additional room within their available space for phage DNA and are considered Defective.
- Defective phage although able to infect respectable bacteria does not have a full complement of phage genes. Such phages are unable to carry a complete cycle in infected bacteria. But they infect recipient cell and insert donor bacterial DNA.
- This donor DNA is incorporated into recipient chromosome.

Transuding phage particle in lysate contains only small portion of entire donor chromosome, the possibility of donor gene transferred to recipient bacteria is quite low and usually one out of 10^5–10^8 recepient cell is transuded for a specific gene.

Specialized transduction

Specific genes are transferred to recipient cell in specialized transduction which occurs only with temperate phage. In specialized transduction phage DNA integrates at a specific attachment site in bacterial chromosome.

- The phage DNA in chromosome is called a prophage and bacterium carrying such prophage is called *lysogenic bacterium*.
- Different prophage may be integrated at specific sites in host DNA, and a host cell may be lysogenizied for more than one prophage at the same time.
- Upon induction with UV light or some other agent, a portion of bacterial genome is removed with some of the phage DNA. The material is replicated and packaged into maturing phage particles, which are released from lysed cell.

- Since the size of DNA molecule incorporated into phage particle remains relatively constant, an equal amount of phage DNA is lost for host DNA that is inserted into phage.
- These progeny bacteriophage lack a complete phage genome and are considered Defective. Defective phage particles can still inject their genetic material into recipient cells, where genetic exchange occurs between donor and recipient DNA.
- The efficiency of gene transfer for specific genes is very high in specialized transduction, because bacteriophage carries only then bacterial genes adjacent to phage at the time of execution.

The most widely used and best understood example of specialized transduction is bacteriophage mediated transfer of galactose (gal) genes, which controls galactose fermentation in *E. coli*. Lambda phage particle infecting *E. coli* have their DNA inserted into chromosome of host at a site between loci for gal and bio. Prophage may detach from genome after induction carrying with it either gal genes or bio genes and enter lytic cycle. These lambda phage particles are either lambda dg (defective carrying gal genes) or lambda dbio (defective carrying bio genes).

The frequency of specialized transduction can be increased through a phenomenon known as high frequency transduction. *Eg:* lambda dg and normal phages is used to infect gal$^-$ *E. coli* at high multiplicity. A high multiplicity is achieved by infecting culture with Defective transducing phage and a large number of helper phage particles. Under such conditions bacteria infected with lambda dg phage are also infected with normal lambda phage and are considered to double lysogens. The lysate after infection with roughly have 50% of transduced lambdaphage.

Abortive transduction

In certain instance of transuding phage DNA in not incorporated to host chromosome. The exogenote neither replicates nor is destroyed; it remains seperate from endogenote and is transmitted unilinearly to one of the two daughter cells formed at cell division. Such abortive transduction is very common and is easily recognized by expression of exogenote in bacteria carrying genetic material.

Eg: Bacteria lacking Proline synthesis (pro$^-$) is infected with transducing phage carrying Proline genes (pro$^+$) and abortive transduction occur. This bacterium is able to grow in medium containing no Proline as they synthesize Proline. As bacteria divides exogene is carried to one or two generations or until the exogenote is not expressed or until the endogenous pool of Proline carried over from parent cell is depleted. This pattern of unilinear transmission of exogenote continues in subsequent cell division. This resell in the formation of slow growing colonies of daughter cells (as microcolonies) lacking exogenote.

Transformation

Transformation is uptake by a cell of naked DNA molecule or fragment from the medium and incorporation of this molecule into recipient chromosome. In natural transformation DNA come from donor bacterium. The process is random and any portion of genome may be transferred between bacteria.

When bacteria lyse, they release considerable amount of DNA into surrounding environment. These fragment may be relatively large and contain several genes. If a fragment contacts a competent cell; one able to take up DNA and be transformed, it can be bound to cell and taken inside. The transformation frequency of very competent cells is around 10^{-3} for most genera *i.e.,* one cell in every 1000 will take up the DNA.

Competency is complex phenomenon and is dependent on several conditions

- Bacteria need to be at certain stage of growth, as exponential phase in *Streptococcus pneumonia*
- When population reaches 10^7–10^8 cell/ml
- When population become competent, they secrete small proteins called competence factor and stimulate the production of new protein (8–10 proteins)

Natural transformation has been discovered in certain gram portion and gram negative gene - *Streptococcus, Bacillus, Haemophilus, Helicobacter, Pseudomonas.*

The mechanism of transformation is extensively studied as in *Streptococcus pneumonia*

- A competent cell binds double stranded DNA fragment (the process is random)
- DNA is then cleaved by endonuclease to double stranded fragment about 5–15 kilobase in length.

- One strand is hydrolyzed by an envelope –associated exonuclease during uptake, other stand associate with small proteins means through plasma membrane
- Single stranded fragment can then align with a homologous region of genome and get integrated

Haemophilus like genes will not produce competence factors but take up DNA from only closely related species (*S. pneumoniae*) and is less particular about source of DNA. The speciality of *Haemophilus* transfer is, due to 11 bp sequence (5'AA47CCC-CA3) that is repeated over 1,400 times in *H. influencae* DNA, DNA must have this sequence to be bound by competent cell.

Artificial transformation is carried out in laboratory by various method, including treatment with $CaCl_2$ (this method succeeds even with species that are not naturally competent as *E. coli*).

Transposons

Transfer of DNA and mutation were long considered to be only mechanism for modification of genetic sequence in living organisms. In 1940s however Barbara Mc Clintok discovered a new mechanism for chromosomal modification. Mc Clintok in her studies on pigmentation in corn plant zea mays noted a variation that was caused not by mutation, but by the movement of discrete genetic elements among chromosomes. It was until two decades later in 1960s that similar controlling genetic elements were discovered and characterized in *E. coli*. These transposable elements or transposons move blocks of genetic material from one location to other on other bacterial chromosome, plasmid or bacteriophage.

Transposons are now recognized to be common constituents of chromosomes, plasmids and bacteriophage. They contain genes coding for antibiotic resistance, carbohydrate utilization, toxin production, amino acid synthesis and hydrocarbon degradation. Electron micrographs have revealed that ends of transforms contain inverted repeated base sequence of 98–40 bases (it is believed that these inverted repeating ends recognize regions in DNA for insertion- possibly those DNA regions that have base sequence complementary to repeat sequence).

Early genetic elements indicated that genomic instability may be caused by transposable elements. The first indication that genomes may contain unstable and possibly mobile comments was the appearance of unexpected phenotypes among progeny of certain strains of maize.

Barbara Mc Clintok's Experiment

It had been known for some time that variegation in color of observed in Indian corn was caused by unstable mutation. The color of the kernel was due to factor encoded by maize C locus. When a gene is mutated no purple pigment is made and kernel appears almost white. The spotted kernel shows the result of reversion in some of kernel cells. Where ever the mutation has reverted, the reverted cells and its progeny will be able to make pigment giving rise to a dark spot on the kernel. It was striking that so many spots were present in kernel. That means mutation is very unstable, it reverts at a rate much higher than expected on ordinary mutation.

Bacterial Transposons

1. *IS Elements*

The simplest bacterial transposons are Insertion sequence or Is elements. An IS elements is a short piece of DNA less than 2000 bp long, that can be inserted at a specific site in chromosome or plasmid. Unlike transposons which are longer and contain genes that are unrelated to insertion as well as insertion genes IS's contain only insertion genes. IS's can be instrumental in regulating the expression of certain phenotypes.

Eg: Salmonella (Gram negative flagellated) bacteria can synthesis two types of flagellins H1 and H2; are located in operons residing on different chromosomal locations. The H2 gene is closely linked to another genes RH1, that codes for repressor of H1 operon. Thus when H2 operon is turned on repressor is synthesized and represses H1 operon. The promoter for H2 operon is located within an IS. This IS contains a gene, 'hin' with a product that is a protein mediating the inversion of entire IS and its site specific recombination with DNA. When IS is in normal orientation, H2 gene is transcribed and H1 operon is repressed. However if IS is inverted with Promoter facing in operate direction of H2 operon, H2 gene (including RH1) is not transferred. The absence of repressor results in transcription of H1 gene. Thus orientation of IS determines type of flagellin synthesized by *Salmonella*.

Regulation of Flagellum synthesis by IS sequence is as follows:-

A. IS containing hin gene is oriented so that flagellin H2 gene and RH1 genes are transcribed. RH2 codes for repressor that repress H1 operon.

B. IS sequence is inverted with Promoter facing in opposite direction of H2 operon. Under such condition H2 and RH1 gene are not transcribed.

Although various types of IS elements are structurally distinct from one another, they show many common features.

- IS elements are boarded on their 5'and 3' ends by short inverted terminal repeats. (IRs) typically between 15 and 25 bp in length.
- The two IR's carried by any particular element are identical or nearly identical in sequence, but inverted in direction.
- Between two IR's is an open reading frame that encodes functions necessary for IS elements, transportation or movement from one chromosomal location to other.
- IS sequence are relatively simple encoding only transposons. These enzyme function to excise IS elements from existing chromosomal site and splice it into new chromosomal portion within the host chromosome.
- Transposons recognize spécific sequence within each of the inverted repeat sequence at the end of encoding elements.
- They make staggered cuts in host DNA sequence adjoining the element. Finally Transposons make cuts at a integration site somewhere within genome and these recombination event is non homologous.
- When elements is integrated into new chromosomal site staggered cuts are subsequently filled in resulting in generation of or short direct repeat of host DNA sequence on either side of interested element.

2. *Composite Transposons (Tn)*

Tn are inserted when two IS elements insert near each other. The insertion sequence are sometimes called 'selfish DNA' *i.e.*, IS replicates at expense of its bacterial host and apparently provides nothing useful in return. An interesting and important aspect of structure of Tn elements is inclusion of IS elements as right and left arms.

In Tn 9 the flanking IS elements are in same orientation with each other whereas Tn 5 and Tn 10, the orientation is inverted. The genes between flanking IS elements confer resistance to antibiotics- Chloramphenicol resistance in Tn 9 kamamycin, bleomycin and streptomycin resistance in Tn 5 and Tetracycline in Tn 10.

3. *Tn 3 Elements*

The elements in this group of transposes larger than IS elements and usually contain genes that are not necessary for transposition (feature that is also characteristic of composite Transposons). Unlike composite transposes Tn 3 elements have simple inverted repeats 38–40 nucletide pairs long at their termine. Tn 3 elements produces target site duplication when they interest into DNA.

The genetic organization of the elements known specifically as Tn 3 are three genes tnp A, tnp B and bla encoding respectively a transposase, repressor or resolvase and an enzyme called beta lactamase. Beta lactamase confer resistance to antibiotic penicillin and other two enzyme play important role in transposition.

The transportation of Tn 3 occurs in two stages

1. Transposase mediates fusion of two molecules forming a structure called co-integrate.
2. Tnp R encoded resolvase mediates a site specific recombination event between two Tn 3 elements.

4. *Conservative Transposons*

- Conservative transposons starts in same way as replicative by nicking and joining strands of donor and target DNA.
- Instead of replication new necks appear in donor DNA on either side of transposon. This releases donor DNA but leaves transposon still bound to target DNA.
- Once remaining nicks are sealed the donor DNA thus lost its transposon where as transposon move towards a target DNA.

Medical Significance of Transposons

Many bacterial transposons carry genes for antibiotic resistance. Resistance thus spreads horizontally and vertically in bacterial population and cells become resistant.

Eg: Staphylococcus, Enterococcus, Neisseria, Shigella and *Salmonella.*

It is conjugative R plasmids which carry resistance. They have 2 components

1. ***Resistance transfer*:** Factor (RTF)- genes needed for conjugative transfer
2. ***R-determinant*:** Genes for antibiotic resistance.

Transposons have great impact on public health. Composite transposons like Tn 10 with multiple antibiotic resistances are an advantage to the bacteria, which harbor it.

Eg: Diary in Illinois' become contaminated with bacteria *S. typhimureum* in 1985 which lead to death of 1400000 people in six states suffering from Salmonellosis.

Transposition

- When TE move they often show little sequence selectivity in their choice of insertion sites. As a result transposons can insert within genes, often completely disrupting gene function.
- They are most common source of mutation in many organisms.
- The mechanism of transportation is called cut and paste transposition (involves excision of transposon from initial location in host DNA, followed by integration of excised transposons to new DNA site).
- To initiate recombination transposase bind to terminal in nerved repeats at the end of transposon. Transposase recognizes the sequence, bring two ends of transposon DNA together to generate Protein-DNA complex called *transposome.*
- After excision of transposome 3'OH end transposon DNA, the ends attach DNA phosphodiester bonds at the site of new insertion.

1.5. IDENTIFICATION OF DNA AS GENETIC MATERIAL

Transformation Experiments

The development of current idea that DNA is the genetic material began with an observation in 1928 by Fred Griffith, who was studying on the bacterium responsible for human pneumonia *i.e., Streptococcus pneumoniae.*

The virulence of the bacterium was known to be dependent on a surrounding polysaccharide capsule that protects it from defense systems of the body. This capsule also causes the bacterium to produces smooth – edged (S) colonies on agar surface. It was known that mice were normally killed by S bacteria. Griffith isolated a rough edged colony mutant which is proved to be both non-capsulated and non-lethal. With the experiment he conducted, he made a significant observation.

- Both R and heat killed S were non lethal
- A mixture of R and heat killed S was lethal.
- The bacterial isolated from mouse that had died from such a mixed infection were only S *i.e.*, live R and heat killed S was lethal because R had somehow been replaced by or transformed to S bacteria.

Several years later of was shown that mixture of R and heat killed S was grown on culture fluid, to give living S cells. A possible explanation for this surprising phenomenon was that R cells restored the viability of dead S cells, but this ideas was eliminated by observation that living S cells grew even when heat killed S culture in mixture was replaced by cell extract prepared from broken S cells, which had been freed from both intact cells and the capsular polysaccharide by centrifugation. Hence, it was concluded that cell extract contained a transforming principle, the nature of which was unknown.

The next development occurred some 15 years later when Oswald Avery, Colin Mac Leod and Maclyn Mc Carty partially purified transforming principle from the cell extract and demonstrated that it was DNA. These workers modified known schemes for isolating DNA and prepared samples of DNA from S bacteria. They added this DNA to live R bacterial culture, after a period of time they placed a sample of S containing R bacterial culture on an agar surfase and allowed it to grow to form colonies. Some of the coloniess (about 1 in 10^4) that grew were S type. To show that this was a permenant genetic change they dispersed many of the newly formed S colonies and placed them on second surface. The resulting colonies are again S type.

Because S and R colonies differed by a polysaccharide coat around each S bacteria, the ability of purified polysaccharide to transform was also tested, but no transformation was observed.

The evidence that transformation was actually caused by DNA alone include four procedures.

1. Chemical analysis of sample showed that major component of transforming principle was dexoy-ribose containing nucleic acid
2. Physical measurements showed that sample contained a highly viscous substance having the properties of DNA.
3. Experiments demonstrated that transforming activity is not lost by reaction either
   a) A purified proteolytic (protein hydrolyzing) enzymes - trypsin, chymotrypsin or mixure of both.
   b) Ribonuclease (an enzyme that depolymerizes RNA).
   Thus the active principle was neither protein nor RNA
4. Treatment with materials known to contain DNA depolymerizing activity (DNase) inactivated the transforming principle.

   Transformation experiment had little initial impact and people who supported "genes-as-protein" theory posed two alternative explanations for the results- that
   a. the transforming principle might not be DNA, but rather one of the protein invariably contaminating the DNA sample
   b. DNA somehow affected capsule formation directly by acting in metabolic pathways for the biosynthesis of polysaccharide and permanently altering the pathway.

Even though the experiment showed insensitivity to proteolytic enzyme and sensitivity to DNase, however since DNase was not pure enzyme, the possibility could not be eliminated conclusively. The transforming experiment was reported 5 years later by Rollin Hotchkins. It was found that extensive purification did not reduce transforming activity. This result supported view of Avery, Mc Leod and Mc Carty but still did not prove it.

Chemical Experiments

The hypothesis of tetranucleotide structure for DNA arose from belief that DNA contained equimolar quantities of Adenine, Thymine, Guanine and Cytosine. The incorrect conclusion arose for two resons. First, in chemical analysis of DNA, technique used to separate base before identification did not resolve them very well, so that quantitative analysis was poor. Second, DNA analyzed was usually isolated from eukaryotes (and in eukaryotes four base are nearly equimolar) or from bacteria whose DNA happened to have nearly equal amounts

Chargaff used new separation and analytical techniques and showed that molar concentration of base could vary widely. Upon publication of Chargaff's results, the tetranucleotide hypothesis quietly died and DNA-gene idea began to catch on.

Blender Experiment

An elegant confirmation of genetic nature of DNA came from experiment with *E. coli* phage T2. This experiment known as blender experiment, because a kitchen blender was used as a major piece of apparatus, which was performed by Alfred Hershey and Martha Chase, who demonstrated that DNA injected by phage particle into bacterium contain all of information required to synthesize progeny phage particles.

- A single particle of phage T2 consist of DNA, encased in a protein shell. The DNA is only Phosphorous containing substance in phage particle.
- The proteins of the shell which contain amino acid methionine and cysteine have only Sulfur atoms.
- By growing phages in nutrient medium in which radioactive Phosphate ($^{32}PO_4^{3-}$) is the sole source of Phosphorous, phage containing radioactive DNA can be prepared.
- If growth medium contains radioactive Sulfur ($^{35}SO_4^{2-}$), phage containing radioactive protein is obtained.
- If these two kinds of labeled phages are used in an infection of bacterial host, phage DNA and protein molecules can always be located by their radioactively

Hershey and chase used phage of this kind to show that P but not S is injected into bacterium.

Each phage T2 particle has a long tail by which if attaches to sensitive bacteria. Hershey and Chase showed that attached phages can be torn from a bacterial cell well by violent agitation of infected cells in a kitchen blender. Thus it was possible to separate an adsorbed phage from bacterium and determine the component (S) of phage that could not be shaken free by agitation— presumably those components had been injected into the bacterium.

- In the first experiment ^{35}S labeled phage particles were adsorbed to bacteria for few minutes.

- Bacteria were then separated from unadsorbed phage and phage fragments by centrifuging the mixture and collecting the sediment (pellet), which consisted of phage-bacterium complexes. Then complexes were re-suspended in liquid and blended.
- Suspension was again centrifuged and pellet which now consisted almost entirely of bacteria and supernatant was collected. It was found that 80% of ^{35}S was in supernatant and 20% were in pellet. (The 20% of ^{35}S remains associated with bacteria was shown many years later to consist mostly of phage tail fragments that adhered too tightly to bacterial surface to be removed by blending)
- A different result was obtained when phage population was labeled with ^{32}P. In this case 70% of ^{32}P remained associated with bacteria in pellet and only 30% was in supernatant.
- Of the radioactivity in supernatant, roughly one third could be accounted for by breakage of bacteria during blending (The remainder was shown some years later to be result of defective phage particle that could not inject their DNA).
- When the pellet was re-suspended in growth medium and re-incubated, it was found to be capable of phage production. Thus ability of bacterium to synthesise progeny phage is associated with transfer of ^{32}P, and hence of DNA, from parental phage to bacteria.

Another series of experiments known as transfer experiments, supported interpretation that genetic material contains ^{32}P but not ^{35}S. In these experiments progeny phage were isolated after blending fem infected cells with either with ^{35}S or ^{32}P containing phage and progeny were then assayed for radioactivity. It was found that no ^{35}S, but about half of the injected ^{32}P was transferred to the progeny. This result indicated that ^{35}S might be residually associated with phage infected bacteria; it was not a part of phage genetic material.

The interpretation (Now known to be correct) of transfer of only ^{32}P was that progeny DNA is selected at random for packaging into protein coats and that all progeny DNA is not successfully packaged.

RNA as Genetic Material

Even through DNA is the genetic material of most organisms including many viruses, some viruses however have RNA as genetic material. In 1958 Sinsheimer showed that genetic material of bacteriophage ØX174 is single stranded DNA. In 1961 Jacob and Monad postulated their presence and function.

1.6. TYPES OF RNA

Three types of cellular RNA have been distinguished- 1) messenger RNA (mRNA) 2) ribosomal RNA (rRNA) and 3) transfer RNA (tRNA). Ribosomal and transfer RNA comprise about 98% of all RNA.

Ribosomal RNA

It is found in ribosome. It comprises about 80% of total RNA of the cell. The base sequence of RNA is complementary to that of the region of DNA where it is synthesized. In prokaryotes the rRNA molecule is formed on a part of DNA strand called ribosomal DNA. In eukaryotes ribosome are formed on nucleolus. The nuclear organizer contains rDNA which transcribes 45S RNA. This then gives rise to 28S RNA through a series of intermediate steps. rRNA is formed from only a small section of DNA molecule and hence there no definite base relationship between rRNA and DNA as a whole.

Depending upon the ionic strength, temperature and pH, RNA molecules may be short compact rod, a compact coil or an extended strand. Ribosomal RNA consists of a single strand twisted upon itself in some regions. The helical regions may show presence or absence of positive interactions. In helical regions most of the base pairs are complementary and are joined by hydrogen bonds. In the unfolded single strand regions the base has no complements. Hence RNA does not show purine-pyramidine equality.

r RNA strands unfold heating and refold upon cooling. The ribosome consists of proteins and RNA. The 70S ribosome of prokaryotes consist of 30S submit and 50S subunit. The 30S subunit consists of 16S r RNA, while 50S consist of 23S and 5S r RNA. The 80S eukaryotic ribosome consists of 40S and 60S subunit. In vertebrates 40S submit contains 18S r RNA, while 60S contain 28S, 5.8S and 5S r RNA. In plants and invertebrates, 40S submit contains 16–18S RNA, while 60S subunit contains 25S and 5S and 5.8S r RNA.

There are 3 types of ribosomal RNA on the basis of sedimentation and molecular weight.

1. High molecular weight rRNA with molecular weight of over a million *e.g.:* 21S – 29S RNA
2. High molecular weight rRNA with molecular weight below a million 12S – 18S rRNA
3. Love molecular weight rRNA – 5S rRNA.

Messenger RNA (mRNA)

Jacob and Monod (1961) proposed the name messenger RNA for RNA carrying information for protein synthesis from DNA (genes) to sites of protein formation (ribosomes). It consists of only 3.5% of total cellular RNA. The molecular weight of an average sized mRNA molecule is about 500,000 and its sedimentation coefficient is 8S. But mRNA varies greatly in length and molecular weight. Since most proteins contain at least 100 amino acid residues, mRNA must have at least 100x3=300 nucleotides on the basis of triplet code.

Messenger RNA is always single stranded. It contains mostly bases Adenine, Guanine, Cytosine and Uracil. There are few unusual substituted bases. Although there is a certain amount of random coiling in extracted mRNA, there is no base pairing (in fact base pairing in MRNA strand destroys its biological activity).

Since mRNA is transcribed on DNA (genes), its base sequence is complimentary to that segment of DNA on which it is transcribed. Usually each gene transcribes its own mRNA. Therefore there are approximately as many types of mRNA molecule as there are genes. There may be 1000–10,000 different species of mRNA in a cell. These mRNA types differ only in sequence of their base and length. When one gene codes for a single mRNA strand, mRNA is said to be monocistronic. When several adjacent genes (cistrons) may transcribe an mRNA molecule, it is said to be polycistronic or polygene.

mRNA molecule has the following structural features.

1. ***Cap*:** At 5’ end of mRNA molecule in most eukaryote cells and animal virus molecule is found a cap. This is blocked methylated structure. The rate of protein synthesis depends upon presence of the cap. Without the cap m RNA molecules bind very poorly to ribosome’s.
2. ***Non-coding region*:** The cap is followed by region of 10–100 nucleotides. This region is rich in A and U residues, and does not translate protein.
3. ***Initiation codon*:** Initiation codon is AUG in both pro and eukaryotes.
4. ***Coding regions*:** Consist of 1,500 nucleotide on the average and translate proteins.

5. ***Termination codon*:** Termination of translation on m RNA is brought about by termination codon UAA, UAG or UGA.
6. ***Non coding region 2*:** Consist of 50–150 nucleotides and does not translate protein. This region contains AAUAAA sequence.
7. ***Poly (A) sequence*:** At 3' end there is polyadenylate or poly (A) sequence which initially consist of 200–250 nucleotides, (becomes shorter with age). The poly (A) sequence is added in the nucleus before mRNA reaches cytoplasm.

The cell does not contain large quantities of m RNA. This is because mRNA unlike other RNA's is constantly underlying breakdown. It is broken down in to consistent ribonucleotides by ribonucleases (in bacteria, mRNA at one end translate proteins while other end undergo breakdown)

Hetrogeneous Nuclear RNA (hnRNA)

Eukaryotic cells synthesize heterogeneous nuclear RNA considered as precursor of mRNA. Only some mRNA serves as precursor, most synthesized are rapidly degraded without becoming mRNA.

Transfer RNA (tRNA) or Soluble RNA (sRNA)

10–20% of total RNA constitutes tRNA. After rRNA second most common RNA is transfer RNA. It is also called soluble RNA because its too small to be precipitated by ultracentrifugation at 100,000 g. It is made up of 73.93 nucleotides (molecular weight 25000–3000). The function of tRNA is to carry amino acid to mRNA during protein synthesis. Each amino acid is carried by a specific tRNA. Since 20 amino acids are carried by specific tRNA coded to form proteins; there must be at least 20 types of RNA. However there are more tRNA molecules than amino acid types (At least two types of tRNA for each amino acid).

tRNA is synthesized in the nucleus of DNA template. Only 0.025% DNA codes for tRNA. Therefore it does not show any obvious base relationship to DNA.

tRNA consist of single stranded loops about itself. The 3' and always terminate in CCA sequence. The 5' end terminates in G or C sequence. Several models for secondary structure of tRNA have been proposed, and of these clover leaf model of Holley *et al.* (1965) is most widely accepted.

Single polynucleotide chain of tRNA is folded upon itself to form 5 arms. As a result of folding the 3' and 5' ends of chain come near each other. An arm consists of stem and a loop. In double helical stem there is internal Watson Crick base pairing which follows A-U and G-C combinations (except in an occasional G-U base pair or mismatch). There is no base pairing in the loops. One of the arms has a stem but not a loop and is called acceptor stem. The other arms are called D arm, the anticodon arm, the variable arm and TpC arm. The variable arm may or may not have a stem. The acceptor stem consist of 7 base pairs and 4 unpaired nucleotide units. The latter include a constant 3' terminal CCA sequence and a fourth nucleotide which is variable purine (A or G). The amino acid molecule attaches to 3' terminal of CCA sequence which is known as amino acid binding site. The 5'end of tRNA is either Guanine (G) or cytosine (C).

The second arm is called D arm. It consists of 15–18 nucleotides with 3–4 base pairs in stem and 7–11 unpaired nucleotides in loop. The loop of D arm is called dehydrouridine (DHU) loop or D loop. It contains two variable regions α and β on either side of two constant Guanine residues.

The third arm anticodon arm consists of an anticodon stem of 5 base pairs and loop called anticodon loop. This loop consists of 7 unpaired nucleotide of which middle three forms anticodon. The anti codon recognizes complementary bases which constitute the codon of mRNA

The variable arm is of two types. In one type loop contain 4–5 bases but no stem, in other type arm consist of 13–21 residence and both stem and loop can be distinguished.

The TpC arm consist of stem having 5 base pairs and a loop of 7 nucleotide the outermost of 5 pairs of stem is C-G

***Unusual bases*:** In addition to usual bases A, U, G and C, tRNA contain a number of unusual bases and in this respect differ from mRNA and rRNA. The unusual bases of RNA account for 15–20% of total RNA of cell. The unusual bases of tRNA are formed by methylation (addition of CH_3 or methyl group to unusual bases *Eg:* Cytosine and Guanine on methylation yield methyl Cytosine and methyl Guanine respectively). The unusual bases are important because they protect tRNA molecule against degradation by RNase. This protection is necessary because RNA is found floating freely in the cell. The structure includes:

1. ***The acceptor arm*:** Which is formed by base pairing between nucleotide of 5' and 3' ends of tRNA. The sequence CCA, which occurs of 3' terminus is not base paired and is the point of attachment for amino acids.
2. ***D or DHU arm*:** Step loop structure containing dehydrouracil, an unusual pyramidine nucleotide.
3. ***The anticodon arm*:** Responsible for recognizing binding codons in mRNA.
3. ***The extra optional or variable arm*:** Occurs only in some tRNA. It may be small containing 2–3 nucleotides or larger congaing 13–21 nucleotide with up to five bare pairs in stem.
4. ***The TpC arm*:** Contains the sequence TpC where p is modified nucleotide called Pseudouracil.

1.7. TYPES OF DNA

The structure identified by Watson and Crick is known as B form of DNA and is believed to be he idealized form of structure adopted by virtually all DNA *in vivo*. It is characterized by a helical repeat of 10 bp/turn by the presence of base pairs lying on the helix axis and almost perpendicular to it and by having well defined major and minor groves.

DNA can be induced to form an alternative helix known as A form under conditions of low humidity. A form is right handed (like B form) but has wider more compressed structure in which base pairs are titled with respect to helix axis and actually lies off his axis. The helical repeat of A is around 11 bp/turn. It is unlikely that native DNA occurs in A form, but DNA-RNA hetero duplexes (double helices having one DNA and one RNA strand) and RNA double helices occur in this form *in vivo*.

A further unsual helical structure can be formed by DNA. The left handed Z-DNA, contains 12 bp/turn and has a pitch 69 A° in Z-DNA, the sugar phosphates backbone follows a Zig–zagged path giving it the name Z-DNA or Z-form. Z-DNA has a single groove with a greater density of negative charges than either major or minor groove of B-DNA. The Z-DNA was discovered *in vitro* (as stable synthetic double stranded DNA consisting of purely alternating pryamidine- purine sequences such as 5' CGCGCG 3')

C DNA is a right handed helix having 9–33 bp/turn, 19A° diameter and 31A° pitch. It is more tightly wound than B from DNA.

1.8. DNA REPLICATION

Genetic information is transferred from parent to progeny organisms by a faithful replication of parent DNA molecules

One of the most important properties of DNA is that it can make exact copies of itself. This process is called *replication* and is the very basis of life. The two strands of DNA double helix are united by hydrogen bonds between purine and pyramidine base pairs. When hydrogen bonds break, two strands separate and unwind. The nucleotide contains free nucleotides which form the nucleotide pool. The nucleotide included in the pool contains Adenine, Guanine, Cytosine and Thymine nitrogenous bases. The free nucleotides pair with nucleotide of two separated strands by means of hydrogen bonds; free Adenine nucleotide pairs with Thymine nucleotide of the strands and free Guanine with Cytosine nucleotide of strands (A-T and G-C pairing). In this way new strand is formed around each old strand- Thus the result of replication is the formation two double strands each identical to the original double helix.

Basic rule for Replication of All Nucleic Acid

1. Nucleotide monomers are added one by one to the end of growing strand by an enzyme called DNA polymerase.
2. The sequence of bases in each row or daughter strands is complementary to the base sequence in old or parent strand being copied *i.e.*, if there is an Adenine in the parent strand a thymine will be added.

Outline of Replication

1. Replication is a semi conservative process in which each of the two double halves from parent double strands have one old and one new strand (repair replication is non conservative).
2. DNA replication machinery
   a. DNA Template
   b. A primer
   c. Deoxy-ribonucleotide triphosphate (dATP GTP, dTTP and TP)
   d. Mg^{++}
   e. DNA unwinding protein
   f. Super helix relaxing protein

g. Modified RNA polymerase (to synthesize RNA primer)

h. Ligase (joining enzyme)

3. Replication starts at a specific point called the origin
4. According to one model, replication starts with a nick or incision made by incision enzyme (endonuclease). As replication starts with incision made by endonuclease the two strands of DNA double helix unwind with the help of DNA unwinding protein which binds to single DNA strands
6. The unwinding of strands impose strain which is relieved by the action of superhelix relaxing protein
8. Deoxy ribonucleotides are added to the 3' end of RNA primer and main DNA strand is synthesized on DNA template. This strand is complementary to DNA strand and is catalyzed by DNA polymerase III
9. The enzyme DNA polymerase I degrades the RNA primers and simultaneously catalyses the synthesis of short DNA segment to replace the primer. This segment is then attached to main DNA strand by DNA Ligase
10. Replication take place discontinuously in other strand and short pieces called Okazaki fragments are synthesized

    Thus, one strand may synthesize a continuous strand and other Okazaki fragments or both strands may synthesize Okazaki fragments. Both new strands are in 5' to 3' direction and one strand is synthesized forward and other backwards.
11. The Okazaki pieces are joined by polynucleotide Ligase an enzyme which joins to from continuous strands
12. Replication may be in one direction (Unidirectional) from the point of origin or in both directions (Bidirectional as in prokaryotic DNA).

Semi Conservative Replication of Double Stranded DNA

According to semi conversation mode proposed by Watson and Crick, each strand of two double helix formed would have one old and one new strand.

Although the mechanism of DNA replication is similar in all organisms, the overall process varies depending on the nature of DNA molecules being copied. The simplest and most common form of replication for circular DNA involves a single Origin of replication from which two replication forks eventually meet and fuse and replication terminates. The detailed steps involved in the process are as follows:-

Replication starts with incision

The DNA consists of many replication units and during replication the incision enzyme (endonuclease) cut DNA at specific initiation point.

Origin of replication

Replication starts at a unique site on the chromosome called Origin of replication.

Unwinding of the strand

The DNA double helix unwinds in order to separate two starts for replication. The unwinding of DNA double helix is brought about by DNA unwinding proteins. These proteins bind preferentially to single strands of DNA and promote unwinding double helical DNA. The region of double helix where the proteins bind form Y shaped replication fork.

Super helix relaxing protein

In circular DNA, unwinding imposes a strain. This strain can be relived to some extend by super twisting in unreplicated part of DNA (This protein has nicking-closing enzymatic activity). By introducing nicks in non replicating region of DNA one strand is made to rotate upon other thus reliving the strain. The break is then closed by ligation and replication fork moves on

Template DNA

The previously existing DNA (Parent DNA) on which new DNA is synthesized is called Template DNA. Template DNA act as a master copy on which new DNA is synthesized by Watson Crick base pairing (A-T, G-C). The newly synthesized DNA has the same base pairs composition as parent double helical template DNA.

RNA Primer

Initiation of DNA synthesis requires an RNA primer. DNA polymerases cannot initiate DNA chains *de novo*, the primer RNA is short polynucleotide chain with 5' phosphate end and 3'OH end.

(The RNA primer is synthesized by DNA template close to origin of replication, this synthesis is probably catalyzed by an RNA polymerase)

Chain elongation

Synthesis of new DNA strand take place by addition of DNA nucleotides to 3'OH group of last nucleotides of RNA primer. This synthesis take place in 5'→3' direction and is catalyzed by DNA polymerase III. The newly synthesized DNA chains have primer RNA attached to this 5 ends the RNA primer is hydrolyzed by 5'→3' exonuclease activity of DNA polymerase I. The resulting gaps are filled in by DNA nucleotides again by catalytic activity of DNA polymerase I. The newly made DNA is joined by DNA Ligase to from high molecular weight DNA.

Replication forks

At a point where two strands are separated, a replication fork is formed (the fork appears Y shaped or in bidirectional replication more precisely the forks appear as bubble or an eye).

Okazaki fragment

The duplication of DNA strands is brought about by movement of replication forks. One strand, the leading strand, undergoes a continuous replication process since its template strand has 3' to 5' directionality, allowing the polymerase assembling the leading strand to follow the replication fork without interruption. The lagging strand, however, cannot be created in a continuous fashion because its template strand has 5' to 3' directionality, which means the polymerase, must work backwards from the replication fork. This causes periodic breaks in the process of creating the lagging strand. The primase and polymerase move in the opposite direction of the fork, so the enzymes must repeatedly stop and start again while the DNA helicase breaks the strands apart. Once the fragments are made, DNA ligase connects them into a single, continuous strand. Thus replication of two parental strands take place as simultaneous when the fork moves along as suggested by Okazaki *et al.* (1968). The important components of the Replication mechinaery are explained in Table 1 where as the terminologies used in Replication process are as follows:

***Helicases*:** Enzymes for helix unwinding

Table 1: Enzymology of DNA replication

DNA Helicase	Also known as helix destabilizing enzyme. Helicase separates the two strands of DNA at the Replication Fork behind the topoisomerase.
DNA Polymerase	The enzyme responsible for catalyzing the addition of nucleotide substrates to DNA in the 5' to 3' direction during DNA replication. Also performs proof-reading and error correction. There exist many different types of DNA Polymerase, each of which performs different functions in different types of cells.
DNA clamp	A protein which prevents elongating DNA polymerases from dissociating from the DNA parent strand.
Single-Strand Binding (SSB) Proteins	Bind to ssDNA and prevent the DNA double helix from re-annealing after DNA helicase unwinds it, thus maintaining the strand separation, and facilitating the synthesis of the nascent strand.
Topoisomerase	Relaxes the DNA from its super-coiled nature.
DNA Gyrase	Relieves strain of unwinding by DNA helicase; this is a specific type of topoisomerase
DNA Ligase	Re-anneals the semi-conservative strands and joins Okazaki Fragments of the lagging strand.
Primase	Provides a starting point of RNA (or DNA) for DNA polymerase to begin synthesis of the new DNA strand.
Telomerase	Lengthens telomeric DNA by adding repetitive nucleotide sequences to the ends of eukaryotic chromosomes. This allows germ cells and stem cells to avoid the Hayflick limit on cell division.

***SSB or single stranded Binding protein*:** This protein prevents single stranded regions from annealing or from forming double strand from hydrogen bonds.

***Leading Strand*:** Continuously replicating strand

***Lagging Strand*:** Discontinuously replicating strand

1.9. TRANSCRIPTION

Gene expression is accomplished by transfer of genetic information from DNA to RNA molecules and then from RNA to protein molecules. RNA molecules are synthesized by using enzymes called RNA polymerase. The process by which RNA molecules are initiated elongated and terminated is called transcription

Central Dogma

The basic mechanism of protein synthesis is that DNA makes RNA which in turn makes protein

DNA
↓ Replication
DNA
↓ Transcription
RNA
↓ Translation
Protein

Protein synthesis consists of two main events transcription and Translation. Transcription is the copying of a complementary mRNA strand or a DNA strand. The DNA strand unwinds and one of the two strands form an mRNA strand. During translation the genetic information present in mRNA direct the order of specific amino acids to form a polypeptide or protein.

Basic Features of Transcription

- The precursor in synthesis of RNA are four Ribonucleatide 5' Tri Phosphates (NTP's) ATP, GTP, CTP and UTP
- In polymerization reaction a 3' OH group of one nucleotide reacts with 5'triphosphate of second nucleotide (a pyrophosphate is removed during the reaction). This results in the formation of a phosphodiester bond (This is the same reaction that occurs in synthesis of DNA).
- The sequence of bases in RNA molecule is determined by base sequence of DNA. Each base is added to the growing end of RNA chain chosen by its ability to base pair with DNA strand used as template. Thus A-U, G-C pairing is maintained
- DNA molecule being transcribed is double stranded yet, only one strand serve as template
- RNA chain grows 5'à3' direction (as nucleotides are added to 3'OH group of growing chain)
- No primer is required for RNA polymerase (unlike in case of DNA polymerase)

- Only Ribonucleotide 5'triphosphates participate in RNA synthesis and first base complementary to respective base is laid down in initiation event. 3'OH group is the point of attachment of subsequent nucleotides and polymerization proceeds in a similar manner.

Transcription

1. ***Template strand of DNA:*** Of the two strands one strand is called sense strand which transcribes mRNA; other complimentary strand is called antisense strand.
2. ***RNA polymerase:*** In most prokaryotes, a single RNA polymerase species transcribes all types of RNA. RNA polymerase core from *E. coli* consists of five subunits: two alpha (α) subunits, a beta (β) subunit, a beta prime subunit (β2), and a small omega (ω) subunit. A sigma (σ) factor binds to the core, forming the holoenzyme. After transcription starts, the factor can unbind and the core enzyme proceed with its work. Eukaryotic and Archaeal RNA polymerases have a similar core structure and work in a similar manner, although they have many extra subunits. The sigma factor initiates transcription of mRNA on DNA template and the core enzyme continues transcription (factor recognizes the start signal on DNA and directs bending of the enzyme to PROMOTER region). PROMOTER consist of conserved sequence which are required for binding of RNA polymerase and initiation of transcription.
3. ***Unwinding of DNA double helix:*** The binding of RNA polymerase to DNA results in local unwinding of DNA double helix. A short segment of DNA molecular opens up and synthesis of mRNA now begins on one of the two strands of its DNA (TRANSCRIPTION BUBBLE FORMATION)
4. ***Elongation of RNA chain:*** Chain elongation take place by addition of activated ribo nucleotide tri phosphates (UTP, ATP, GTP and CTP) to one strand of DNA template.

   For each nucleotide added to the growing RNA chain a pyrophosphate (PPi) is given off; this is rapidly hydrolyzed to inorganic phosphate (Pi). The synthesis of RNA chain is energized by expediture of energy in the form of pyrophosphate (For each nucleotide monomer added to the chain two high energy phosphates are expended)

   DNA + Activated ribonucleotide → Polymerase + RNA Template
   triphosphates

4. ***Dissociation of sigma factor*:** After commencement of chain donation the sigma factor dissociates from core enzyme (It can be used by another polymerase molecule)
5. ***RNA chain termination*:** Termination of RNA chain synthesis appears to be brought about by two types of mechanisms. In the first type termination signal appears to be recognized by DNA itself. RNA polymerases reads an extended poly (A) sequence on DNA and this result in an RNA transcript with a terminal poly (U) sequence (which form HAIRPIN STRUCTURES)

   The second type termination signal involves an additional protein called the rho (ü) factor. The ü factor binds to RNA polymerase and promotes the termination and when transcription has been completed rho dissociate from RNA polymerase –DNA complex.

1.10. TRANSLATION

During translation process, proteins are made by ribosomes on mRNA strand. The main steps in translation are

1. Activation of amino acids
2. Transfer of amino acid to tRNA
3. Initiation of synthesis
4. Elongation of polypeptide chain
5. Chain termination

Activation of Amino Acids

The first step in translation is activation of amino acids. The 20 amino acids commonly found in proteins are first screened to eliminate D isomers. Only L amino acids are taken for protein synthesis. The activation of amino acid takes place through their carboxyl groups. Each amino acid is catalyzed by its own specific activating enzyme called amino acyl tRNA synthetase to form an amino acyl adenylate (aaa) synthetase (Aminoacyl AMP) in presence of ATP.

A high energy acyl bond is formed between the Phosphate of ATP and carboxyl group of amino acid

Aa amino acid + ATP + Aminoacyl+RNA synthesis $\rightarrow$ Aminoacyl adenylate synthetase (Aminoacyl AMP) + Inorganic phosphate

Transfer of Amino Acid to tRNA

The transfer of activated aminoacid to tRNA also specify the activated amino acid is transferred to its specific –RNA. A high energy ester bond is formed between the carboxyl group of amino acid and 3' hydroxyl group of terminal adenosine of tRNA. The aminoacyl adenylate which is attached to activating enzyme (synthetase) reacts with specific tRNA to from an aminoacyl tRNA complex

Aninoacyl adenylate synthetase + tRNA → Aminoacyl + adenosine + RNA monophaspha-tase + Aminoacyl tRNA synthetase (enyzme)

Polypeptide Chain Initiation

Polypeptide chain initiation begins with association of mRNA, a 30S ribosomal subunit and formyl methionine tRNA to from 30S Pre initiation complex. The formation of initiation complex requires GTP and 3 protein factors called IF1, IF2, IF3 in prokaryotes and 6 types eIF2, eIFa, eIFa3, eIFa4 and eIF3 in eukaryotes.

IF3 mediates binding of mRNA to 30 S subunit of ribosome IF, and IF2, enhances binding of initiator tRNA to mRNA.

mRNA+30S + fmet tRNA+GTP+IF1+IF2+IF3 → Pre Initiation Complex

A50S subunit of ribosomal subunit joins 30S Pre initiation complex to form a 70 S initiation complex.

Pre Initiation complex + H_2O + 50S → 70s initiation complex + GDP + Pi

Each ribosome has 2 sites, A site and P site. Amino acyl tRNa occupies A site, while met tRNA (tRNA with initiator amino acid methionine) will be at the at P site.

This occupation requires a molecule of GTP and participation of IF1. The energy of GTP is utilized for formation of peptide linkage. Peptide linkage occurs when COOH group of peptidyl tRNA present at P site reacts with amino group of aminoacyl tRNA at A site. The enzyme peptidyl transferase catalyzes the formation of peptide bond. Later aminoacyl tRNA moves from A site to P site (by IF2).

Formation of Polypeptide Chain

A Specific type of tRNA which has bases complementary to the codon of mRNA linked to a particular amino acid get attached to specific site on mRNA, when its anticodons can form a pairing with first codon of mRNA. Soon after, a second tRNA with its amino acid joins the second site (Triplet on mRNA). The first and second amino acid attached to different tRNA form peptide linkage as both are linked to in RNA *via* anticodons of tRNA. A molecule of water is released during formation of peptide linkage.

$$COOH + NH_2 \rightarrow CONH + H_2O$$

As soon as peptide linkage is formed the tRNA is made free and comes back to cytoplasm. Here the sequences of bases in tRNA will be identical with base sequence of that part of DNA coding for mRNA.

Chain Elongation

Elongation of chain requires the participation of certain factors called Elongation factors EF. These are EF-TU, EF-TS and EFG in prokaryotes and EF1 and EF2 in eukaryotes

After the first amino acid tRNA complex has moved from A site of ribosome to P site , a second amino acyl tRNA complex occupies A site. The codons are recognized enzymatically (This requires participation of EF-TU, GTP and Mg^{++})

Peptide linkage take place between 2 amino acids and aminoacyl tRNA complex migrates to P site rendering A site vacant. Now a third amino acid-tRNA complex occupies A site and process continues. The shifting of amino acyl tRNA complex from A site is called translocation. During translocation, ribosome moves relatively to mRNA in 5' → 3' direction.

Polypeptide Chain Termination

Amino-acyl tRNA does not bind to A site of ribosome if the codon is UAA, UAG and UGA. The stop signals are recognized by proteins called release factors

In prokaryotes there are three RF, RF1, RF2 and RF3 and it is RF in eukaryotes

The RF interact with enzyme peptidyl transferase resulting in hydrolysis of bond between polypeptide chain and tRNA (Linked to the last amino acid of peptide chain at site) and polypeptide chain is released from ribosome

Summary of the Events

1. Initiation-Initiator tRNA f met pairs with initiator codon AUG
2. 50S ribosome subunit joins with 30S subunit to form 70S ribosome unit. This completes initiation complex
3. Elongation- aminoacyl tRNA binds to complimentary codon (GUC to CAG) at A site of the ribosome. A peptide bond is formed between f met and valine and ester bond between tRNA in P site and its amino acids (f met) is cleaved
4. Peptidyl tRNA on A site is transferred to P site on ribosome and displaces free tRNA in P site. A site is now free to receive new amino acyl tRNA and cycle of chain elongation is repeated
5. Termination-when ribosome reaches a nonsense codon (UAA) on mRNA, protein synthesis is terminated
6. The protein is released and ribosome is dissociated

1.11. GENETIC CODE

Genetic code is the communication between triplets (three nucleotides) in DNA or RNA and the amino acid in protein. Any one of the possible four nucleotides A, T, G and C can occupy each of the three positions of the codon, so that 64 possible combinations of tri nucleotide sequence exist.

DNA contains four kinds of nucleotide (A, T, G and C) and protein is synthesized from 20 different types of amino acids. A triplet or three letter code was suggested by Physicist Gamou in 1954. According to the triplet code, three letters or bases specify one amino acid; thus 64 (4×4×4) distinct triplets or purine and/or pyramidine bases determine 20 amino acids. These triplets have been called codons. Since there are 64 codons and only 20 amino acids, it is obvious that there are many more codons than are amino acids *i.e.*, the code is degenerate. Experimental evidence shows that code is triplet, one and 61 of 64 codon code for individual amino acids during protein synthesis.

The Code is Non Over Lapping

Since DNA molecule is long chain of nucleotides, it could be read either in an overlapping or non overlapping manner.

In non overlapping code six nucleotides would code for two amino acids, while in over lapping code up to 4 could be coded. In the non overlapping code, each letter is read only once while in over lapping code it would be read three times (each time as a part of different word)

Studies on gene mutations show that the code is of non overlapping type

Code Has Polarity

If gene is to specify the same protein repeatedly, it is essential that the code must be read between fixed start end points. These points are initiation and termination codons respectively. It is also essential that code must be read in fixed direction *i.e.*, code must have polarity.

Wobble Hypothesis

The triplet code is a degenerate one with many more codens than the member of amino acid types could. An explanation for this degeneracy is provided by Wobble typothesis proposed by Crick

Since there are 61 codes specifying amino acids the cell should contain 61 different tRNA molecules each with a different codon. However the number of tRNA molecule types discovered are much less than 61. This implies that anticodons of some tRNA read more than one codon on mRNA

According to wobble hypothesis only the first two positions of triplet codons on mRNA have precise pairing with the bases of tRNA anticodon. The pairing of third position bases of codon may be ambiguous and varies according to the nucleotide present in this position. Thus a single tRNA type is able to recognize two or more codons differing only in this third base.

Eg: Anti coden UCG of serine tRNA recognizes two codons AGC and AGC the bonding between UCG and AGC follows Watson Crick pairing pattern

The Code is Universal

The genetic code is valid for all organisms ranging from bacteria to man. It is essentially the same for all organisms and is therefore said to be universal. The universality of code was demonstrated by Marshall, Caskey and Nirenberg.

The code has remained constant since the time it was fixed when complex bacteria evolved. Any mutation altering the code reading would change the reading of mRNA. This in turn may change the amino acid sequence of protein synthesized by organism. As many of these changes could be lethal, there would be a strong selection pressure against such mutation. Changes in proteins take place only with respect to the particular amino acids. Moreover only a few such changes take place at a time. Most mutations result in single amino acid substitutions.

1.12. REGULATION OF GENE EXPRESSION

The synthesis of an enzyme by a cell in general is dependent to a great extent on the external environment in which the organism is growing. All enzymes required by a cell are not synthesized in equal amounts, some are present is larger amounts than others (Hence it is understood that cell is able to regulate enzyme synthesis). Two simple forms of regulation namely enzyme repression and enzyme induction govern gene expression in a variety of operons in bacteria.

Enzyme Repression

According to the concept of repression, the enzyme catalyzing the synthesis of a specific product is not synthesized if that particular product is present in the media.

Eg: Enzymes involved in the formation of amino acid Arginine is synthesized only when Arginine is not present in culture media.

From this it can be concluded that external Arginine represses the synthesis of these enzymes. The phenomenon of enzyme repression is very wide spread in bacteria and is of great value as of ensure the organism does not waste energy synthesizing unwanted enzymes.

Enzyme Induction

This is a phenomenon complementary to repression where the synthesis of an enzyme occurs only when its substrate is present in the culture medium.

The substance that initiates enzyme induction is called an indicer (the substance that represses the enzyme production is called co-repressor). These are small molecules in nature and are collectively known as effectors.

Eg: Enzyme β galactosidase which is involved in the utilization of sugar lactose is produced if lactose is absent in the medium.

Operons

An Operon is a complete unit of gene expression on gene, generally involving genes coding for several polypeptides on a ploygenic mRNA or gene coding for ribosomal RNA. A specific region of DNA known as operator gene present adjacent to the first gene in operon controls the transcription for that operon.

Jacob and Monad postulated the Lac operon concept which is a model for understanding gene regulation.

Three enzymes are synthesized by *E. coli* for the metabolism of lactose.

1. β galactosidase
2. Galactosidase permease
3. Thiogalactoside transacetylase

The genes that code for these enzymes occur in continuous segment of DNA (polycistron). These genes are appreciably transcribed only in the presence of an inducer.

The lac operon consist of

1. ***Promoter region:*** To which RNA polymers bends
2. ***Regulatory region:*** That codes for synthesise of a repressor protein
3. ***Operator region:*** That occurs between promoter and the three genes laz Z, laz Y and lac A which codes for galactosidase permiase and acetylase respectively

The regulation process includes the following events.

- The regulatory gene codes for a repressor protein, which in absence of lactose bind to operator region of DNA.
- The binding of repressor blocks transcription of all three genes for lactose metabolism
- The transcription is repressed in the presence of inducer, that is lactose. The transcription is repressed lactose metabolism proceeds.
- The binding of lactose with the repressor, repress or inactivates repressor and become enable to bind to operator gene.

Catabolite Repression

It occurs when organisms is offered a catabolizable energy source (lactose) in the presence of more readily calabolizable energy source such as glucose. One consequence of this phenomenon is that can lead to diauxic growth, where if two energy sources are present in the medium at the same time the enzyme needed for utilization of one energy source is synthesized and other repressed.

Eg: Culture medium with a mixture of glucose and lactose; The enzyme b galactosidase responsible for utilization of lactose is subjected to catabolic repression as long as glucose is present in the medium.When glucose is exhausted catabolite repressor is abolished and enzyme is induced.

Attenuation

It is a type of regulation which controls the biosynthesis of amino acids. The least studied is tryptophan operon. The typtophan operon contains

1. Structural genes for five proteins trp A,B,C,D,E of tryptophan biosynthetic pathway
2. Regulatory gene at the beginning of operon
3. Promoter gene
4. Operator gene
5. Additional sequence called leader sequence

The leader sequence contains a region called atternuator which codes of a polypeptide that contains tandem tryptophan units (- trp – trp -) near its terminus. The leader sequence is synthesized if tryptophan is plenty, hence attenuating further synthsis.

The rate at which transcription of trp mRNA begins is controlled by a tryptophan activated repressor molecule. This molecule blocks the access of RNA polymerase to trp promoter.

The phenomenon of attenuation depends on exact folding pattern of nascent mRNA, between the transcribing RNA polymerase and ribosome translating trp L region. The leader protein contains stretches of sequence that can base pair with each other forming hairpin like structure.

When tryptophan is in abundance the ribosome translates freely though the leader and nascent mRNA folds into a structure that signals RNA polymers to terminate transcription.

When tryptophan is scarce, the number of trp activated RNA's are less. Hence ribosome stalls at UGG codons in leader sequence. The mRNA folds itself into an alternative structure, which is not recognized by RNA polymer and as a result proceeds till the end of the operon.

The striking feature of this operon is that synthesis of leader peptide results in termination of transcription of tryptophan structural genes.

Summary of Lac Operon

- Prevention of expression of B-gal gene due to non available of lactose. The repressor forms a tetramer and bind to OPERATOR preventing expression of gene.
- β-gal gene is expressed as lactose is available and act as inducer. The inducer bind to repressor resulting in inactive repressor

Summary of Tryptophan operon:

- Formation of hairpin like structure to signal RNA polymerase to terminate transcription as tryptophan is available in plenty.
- Stalling of the gene occurs as this is not recognized by RNA polymerase therefore indicates scarcity of tryptophan. Ribosome stall at Tryptophan codons, transcription continues in absence of base pairing between hair loop.

1.13. MUTATIONS AND REPAIR

Mutations are a heritable change in the base sequence of DNA of an organism. These changes are largely detrimental, but are of importance is generating new variability and contributing to the process of evolution.

The term mutation was first coined by Hugo de Vreis and is derived from Latin word mutare meaning to change. A strain carrying such a change is called mutant and the process called mutagenesis. A mutant will differ from its parent strain in genotype *i.e.*, genetic composition as there will be a change in the precise sequence of nucleotides in DNA. In addition the visible properties of mutant *i.e.*, phenotype may also show alterations when compared to parental strain.

Types of Mutations

1. Conditional mutations are expressed only under certain environmental conditions.

   Eg: A temperature sensitive mutant which behaves normally below 34°C and as mutant above 39°C. An intermediate state is observed in between these two temperatures.
2. Biochemical mutations inactivate the biosynthetic pathway of organisms and this leads to change in biochemistry of cell. Hence the organism is unable to grow on medium lacking adequate supply of metabolic products. In other words such mutants have lost capacity to grow on minimal medium and therefore requires additional nutrient supplements. The mutants are known as Auxotrophs and the wild type strains capable of growing on minimal media are known as prototroph's.

   Mutations can be feather classified based on numbers of nitric acid base changes in genome. The classification are as follows

Point Mutations

Where only single base is changed either by base subsititutions insertion or base deletion.

Multiple Mutations

Where mutants differ in two or more base pairs from wild type sequences. Often mutations lead to wrong amino acid sequences due to genic changes.

The type of change in amino acid sequence may be taken as criteria for classifying mutations.

1. ***Missense mutations***: A missense mutations is one where altered gene (codon) in mRNA specifies a wrong amino acid resulting in replacement of one amino acid in polypeptide chain by another.

The changed codon may then code to another amino acid.

Such mutations may occur either by substitution deletion or insertion of amino acid coding sequences.

Missense mutations by substitution result in proteins which differ from their normal counterparts only in a single amino acid.

Eg: GAG codes for gulamic acid and when nucleic acid base A is replaced by U, the new codon reads GUG which codes for amino acid valine. Such mutations sometimes lead to production of abnormal hemoglobin

Eg: Sickle cell anemia where sixth amino acid is changed from glutamic acid to valine.

2. ***Nonsense mutations*:** 3 of 64 codons considered as stop codons which do not specify any amino acid (UAA, UGA, & UAG) (They are also known as nonsense codons)

   Nonsense mutation cause early termination of translation and therefore results in shortened polypeptide. This involves the conversion of sense codon to nonsense codon or stop codon. The result in incomplete polypeptide chain synthesized which is likely to be biologically inactive.

   (As polypeptide chain synthesize take place in 5' -à 3' direction a nonsense mutation near 5' end will produce chain with little or no biological activity, whereas near 3' end will produce chain which is near completion which have some or normal biological activity.

3. ***Silent mutations*:** These are a kind of point mutations that cannot be denoted until nucleic acid sequencing techniques are applied and genome is screened. They do not cause any change in phenotypic expression and hence are known as silent mutations.

   Eg: Codon AAG and AAA specify amino acid lysine. If codon AAG undergoes mutation base G is replaced by A and newly formed codon will specify same amino acid.

Leaky Mutations

If an amino acid in protein molecule is replaced by more bulky unrelated amino acid, stereochemical changes as likely to occur. These changes could be manifested as reduction rather than loss of activity of enzyme.

Eg: Bacterium carrying such mutations in an enzyme that synthesize an essential component might grow very slowly, but not completely retarded unless the component is provided in growth medium.

Non-leaky Mutations

These are mutations where changes in substrate binding site of enzyme synthesized are altered.

Eg: Hydrogen bonding to non hydrogen bonding, changes in sulfhydril to other side chains etc.

Base Pairing Substitutions/Tautomerism

Tautomerism is a phenomenon where original keto form of base is replaced by its imino or enol form.

These tautomeric shifts change hydrogen bonding characteristics of bases allowing purine or pyramidine substitutions.

Base pair substitutions are categorized into 2 major types.

1. ***Transitions***: Transitions are most common type of substitutions where purine base is replaced by another purine and pyramidine by another pyramidine.
2. ***Transversions***: Are substitutions where purine base is substituted by pyramidine base or *vice versa* (This type of substitution is very rare)

Changes in Base Pairing

1. ***Insertion***: Insertion is a process where bases are added to genetic code. This insertion of base disturbs the code.

   Eg: CAT **GAT** CAT

   When G is inserted after GAT the code will be read as CAT GAT GCA
2. ***Deletion***: Is a process where bases are removed or deleted from nucleotide chain.

   Eg: CA**T** GAT TCA

   After deletion of T code will be read as CAG ATT CA
3. ***Frameshift mutation***: A Mutation in which there is deletion or addition of one or a few nucleotide. The name is deserved from the fact that there is a shift in reading frame, either backwards or forwards by one two nucleotides.

Addition or deletion of one is two bases results in new sequence of codes which may code for entirely different amino acids. Thus results in drastic change in protein synthesized. The protein synthesized is unusually nonfunctional (if reading frame shifts by 3 nucleotides, resulting protein is normal except that it may lack one amino acid and may contain an extra amino acid).

Mutagenesis

The process of mutation is referred to as mutagenesis. It occurs in one of the two ways. The overall impact of mutagenesis is represented in Table 2. Mutationns due to mutagenic agents are wide and differ in their mode of action or activity. The major mutagenic effects are as follows:

1. ***Spontaneous mutagenesis*:** This kind of mutagenesis occurs in Nature and arises occasionally in all cells and develops in absence of any added agent. It may also result from errors in DNA replication damage to DNA etc.

Table 2: Table showing mutagenic impacts with mode of action

Mutagen	*Basis of action*	*Mutagenic effect*
Chemical Mutagens		
Ethy methane Sulfate, Methyl sulfate	Alkylation of purine by depurination	G-C = AT transition, Tranversion
Hydroxyl amine	Conversion of cytosine to dihydroxyl amino cytosine Alkylation of guanine	G-C = AT transition
N-Methyl N-nitro N-Nitroso guanidine	Deamination, conversion of cytosine to uracil and adenine to hypoxanthine	G-C = AT transition
Nitrous Acid	Adenine analog cause mispairing with cytosine Thymine analogue cause mispairing with Guanine	G-C = AT transition
Base analogs		
2 amino purine	Pyramiding dimer	Frame shift
5 Bromouracil	Intercalation	Frame shift
Others		
UV light		
Ethidium bromide		

2. ***Induced mutagenesis*:** It results when an organisms is exposed to some physical or chemical agents called Mutagens. The mutagens are classified based on their mode of actions.

a. *Base analogs which are incorporated into DNA instead of normal bases*: The Base analogues are similar to nitrogen bases and can be incorporated into growing polynucleotide chain during replication.

   Eg: 5 Bromouracil (5BU) an analogue of thymine, thus when replaces T, produce AT-GC transitions.

b. *Agents modifying purine and pyramidines*: This substance alters a base that is incorporated in DNA and thereby changes its hydrogen bonding specificit group.y.

   Eg: Nitrous oxide where amino group replaced by hydrogen group (-OH)

c. *Intercalating agents*: Such mutagens produce distortions in DNA

   Eg: Acridine orange

   (It blocks base in template strand and does not permit any base to pair with leading to DNA with deficient base pair)

d. *Nitrous acid*: Nitrous acids deaminate cytosine to produce uracil, which base pairs with adenine resulting in alteration fem GC-AT pair.

e. *Ethyl methane sulphonate (EMS), Ethyl ethane sulphonate (EES) and Diethyl sulphate (DES)*: They are alkylating agents leading to alkylation N –T position of Guanine or N-3 of Adenine

f. *Ethidium bromide*: It has dimensions similar to purine - pyramidine base pair (four rings); intercalates into double helix, causes adjacent base pairs to move apart slightly. It disrupts replication.

g. *Radiations:* Among physical mutagens radiations is most important Types of Radiation are as follows:

   i. *Ionizing radiations*: Ionizing radiations such as X–rays, v-rays etc are high energy radiations which change the atomic structure of substance by causing loss of an electron and formation of ions sometime an electron pair may be moved from an inner to outer orbital shell. This brings about excitation of atom. In this excited state atom is highly reactive and is called free radical. It causes chromosomal breakage which may lead to undesirable structural changes such as transactions and inversions.

   ii. *UV light (Non ionizing radiations)*: UV light of wave length between 200-300 nm produces mutations. The primary mutagenic effects of UV light appear due to production of pyramidine dimmers. Adjacent pyramidine become covalently linked to form a cyclobutane ring. Irradiation of bacterial culture and subsequent extraction of DNA yields 3 possible types of pyramidine dimers.

Thymine	-	Thymine	-	50%
Thymine	-	Cytosine	-	40%
Cytosine	-	Cytosine	-	10%

iii. *X-rays*: X-rays bring about mutations by breaking phosphate linkages in DNA. As a result large number of bases is lost.

DNA Repair

The requirements that DNA sequence of a gene is preserved despite the presence of numerous agents that can mutate DNA, has caused organisms to develop methods which prevent mutation by repairing damage to DNA. The repair mechanisms are complex but essentially three main types occur.

Excision Repair

This is complex and most common form of DNA repair. Enzymes recognizes nucleotides forms nicks and remove (Excise) damaged base leaving a gap. DNA polymerase synthesizes a new strand and replaces the missing base which joins to new DNA by enzyme Ligase.

Mismatch Repair

This system corrects errors introduced during DNA replication by identifying mismatched nucleotides. It does this by distinguishing between parental DNA with correct sequence to daughter strand (by enzymes).

Direct Repair

This is much less common form of DNA repair which involves reversal of structural alterations that occur in nucleotides.

Eg: Photo reactivation

Photo Reactivation

Photo reactivation is an enzymatic cleavage of Thymine dimer activated by visible light (300-600 mm). An enzyme called PHOTO LYASE (PR enzyme) has been involved in repair.

- The enzyme bends specifically to thymine dimer (Cytosine dimmers, but more of thymine dimer)
- If light is absorbed thymine dimer is monomerized
- Photolyase dissociates from now normal pair of thymine

SOS Repair

SOS Repair includes a bypass system that allows DNA chain growth across damaged segments at the cost of fidelity of replication. It is an error prone process; even though intact DNA strands are formed, the strands contain incorrect bases (The principle involved is that survival with mutations is better than no survival at all).

- An important feature of SOS repair is requirement for a functional recA gene
- The rec A protein bind to single stranded DNA tightly (but only very weakly it binds to double stranded DNA). The distortion resulting from pyramidine dimer produces a short stable single stranded region to which Rec A binds.
- When DNA polymerase III encounters a dimer site to which Rec A is bound, Rec A interacts with polymerase and inhibits editing function, thereby avoiding idling response and allowing replication fork to advance.
- DNA polymerase III utilize residual base pairing ability to thymine dimer and put two Adenines in daughter strand.
- The presence of Rec A at dimer site inhibits editing and cause mispaired bases to remain in daughter strand as mutation.

Lex A is a repressor that regulates expansion of SOS repair genes Rec A protein stimulated by single standard DNA is involved in the inactivation of Lex A gene repressor thereby inducing the response. The interaction between Lex A and activated Rec A results in proteolytic cleavage of Lex A. Lex A is no longer able to block transcription and SOS repair genes are thereby induced and expressed. Among the genes that are induced are uvr A B C and D and also umu D and umu C. umu D is cleaved by Rec A proteinase activity and trunctated protein umu D in association with umu C forms DNA polymerase V. DNA synthesise by DNA polymerase V is error prone.

1.14. GENE PROTEIN RELATIONSHIP

One gene one enzyme hypothesis is the idea that genes act through the production of enzyme with each gene responsible for producing a single enzyme that inturn affects a single step in metabolic patway. This concept was proposed by Geroge Beadle and Edward Tatum (1941) in an influential paper on genetic mutations in the mold *Neurospora crassa* and subsequently was dubbed the “one gene one enzyme hypothesis” by their collaborator Norman Horowitz

By early 1950’s advances in biochemical genetics spurred in part by original hypothesis -made the one gene one enzyme hypothesis seem very unlikely. Beginning in 1957 Vernon Ingram and others showed through protein finger printing, the genetic variations in proteins (such as sickle cell hemoglobin), could be limited to differences in just a single polypeptide chain in multimeric protein leading to ‘one gene one polypeptide’ hypothesis instead.

In the early part of twentieth century, scientist knew what genes did, but did not know what they were. Francis Crick, one of the co discoverers of three dimensional double helical structure of DNA was among the first to propose that a gene was a linear sequence of necleotides and that each gene encoded a single protein. Crick called this proposal of sequence hypothesis (1958); other scientist has since referred to it as the genes-on- a- string hypothesis. In Cricks words this hypothesis “assumes that the specificity of a piece of nucleic acid is expressed solely by the sequence of its bases and this sequence is a (simple) code for amino acid sequences of a particular protein”. Crick freely admitted that his hypothesis is a just that a hypothesis “for which proof is completely lacking”. Today scientist revealed a “notion that nucleotide sequence (genes) directly dictates amino acid sequences”, and is known as colinearity.

With colinearity, the number of nucleotides in the gene is proportional to the number of amino acids in a protein *i.e.*, a continuous sequence of nucleotides in DNA codes for continuous sequences of amino acids in protein.

Chapter 2

ENZYMOLOGY

2.1. ENZYMES

Enzymes are biological catalysts. They increase the rate of chemical reactions taking place within living cells without them suffering any overall changes. The reactants of enzyme catalyzed reactions are termed suhstrates and each enzyme is quite specific in character, acting on a particular substrate or substrates to produce a particular product or products.

All enzymes are proteins. However many enzyme protein lacks catalytic activity without the presence of a non protein component called a cofactor. In this case, the enactive protein component of an enzyme is termed the holoenzyme. The cofactor may be an organic molecule, when it is known as coenzyme or it may be a metal ion. Some enzymes bind cofactors more tightly than other when a cofactor, and is bound so tightly that it is difficult to remove without damaging the enzyme, it is sometimes called a prosthetic group.

$$\text{Holoenzyme} \rightarrow \text{Apoenzyme} + \text{Co-enzyme.}$$

Where Holoenzyme is Active enzyme, Apoenzyme is protein part and coenzymes the non protein part.

"Enzymes may be defined as biocatalysts (substance that increase the velocity or rate of chemical reaction without itself undergoing any changes in overall process), synthesized by living cells. They are protein in nature, colloidal and thermolabile in character and specific in their action". In recent years certain non protein enzymes (RNA) have also been identified.

Nomenclature and Classification

In early days enzymes were given names by their discoveries in an arbitrary manner. *Eg:* trypsin, pepsin chymotrypsin convey no information about the function of enzyme.

Enzymes are sometimes considered under two broad categories.

1. ***Intracellular*:** They are functional within cells where they are synthesized
2. ***Extracellular enzymes*:** These enzymes are active outside the cell (all digestive enzymes).

A commission was appointed by the International Union of Biochemistry and its report was published in 1964 and updated in consecutive year's forms the basis of present accepted system.

Enzymes are divided into six major classes; each class on its own represents the general types, of reaction brought about by the enzymes of that class.

1. ***Oxidoreductases*:** Enzymes involved in oxidation – reduction reactions *Eg:* Alcohol dehydrogenase.
2. ***Transferases*:** Enzymes that catalyze the transfer of functional groups. *Eg:* Transferases –Hexokinase, Transaminase
3. ***Hydrolases*:** Enzymes that bring about hydrolysis of various compounds *Eg:* Esterase
4. ***Lyases*:** Enzymes specialized in addition or removal of water, ammonia, CO_2 etc *Eg:* Aldolase, fumarase
5. ***Isomerases*:** Enzymes involved in all isomerization reactions. *Eg:* Triose phosphate isomerase.
6. ***Ligases*:** Enzymes catalyzing the synthetic reactions, where two molecules are joined together and ATP is used. *Eg:* Glutamine synthetase

2.2. FACTORS AFFECTING ENZYME ACTIVITY

The contact between the enzyme and substrate is most essential pre–requisite for enzyme activity. The important factors that influence the velocity of enzyme reactions are discussed here under.

1. ***Concentration of enzymes*:** As the concentration of the enzyme is increased, the velocity of the reaction proportionately increases; provided substrate concentration is unlimited.
2. ***Concentration of substrate*:** Increase in substrate concentration gradually increases the velocity of enzyme reaction within the limited range of substrate levels. At very low concentration of substrate rate of enzymic reaction is slow, but of increases.

At very low concentration of substrate rate of enzymatic reaction is slow, but it increases with the increase in substrate concentration. Finally a point reaches where there is no change in the enzyme activity with the increase in substrate concentration. The initial velocity reaches a maximum value (Vmax) which does not exceeded on further rise in substrate concentration. This saturation effect is exhibited by nearly all enzymes.

Effect of substrate concentration on enzymes activity is indicated by rectangular hyperbolic form of substrate saturation curve obtained by plotting initial velocity V against the concentration of substrate (S). It results from the fact that enzyme is the limiting factor in enzyme substrate reaction because enzyme molecules are far less in number than their substrate molecules. So on gradually increasing the substrate concentration all enzyme molecules are ultimately saturated with substrates and act at its maximum capacity. A further rise in substrate concentrate fails to raise the velocity further due to non availability of free enzymes.

Significance of km

Km is called Michaelis constant; km is defined as molar concentration of substrate at which half maximal velocity is attained.

Km value is not influenced by enzyme concentration or non competitive inhibitors but is changed by competitive and uncompetitive inhibition. Sometimes alteration of pH, temperature and ionic concentration and in some cases of allosteric modulators km value is changed.

The substrate affinity of an enzyme and consequently its km value varies from substrate to substrate. No two enzymes have the same km value. Isoenzymes of the same enzyme differs in their km value and consequently in their affinities for the same substrate. Km indicates the degree of affinity of an enzyme towards its substrate. Lower the km higher will be the substrate affinity and vice versa.

Determination of km value helps to determine the natural substrate from a group of substrate. It determines the rate limiting enzyme of a metabolic pathway.

Inhibition

Most enzymes can be inhibited by specific chemicals. These inhibitor chemicals are of two types.

1. ***Irreversible Inhibitors*:** which combine with or destroy a functional group on enzyme molecule that is necessary for its catalytic activity.
2. ***Reversible Inhibitors*:** which do not cause permanent damage in functional groups and once there inhibitors are removed the enzyme becomes fully active. There are two types of reversible inhibitors – *competitive* and *Non competitive*.

Competative or Substrate Analogue Inhibition

Competitive inhibition is a type of reversible inhibition of enzyme. The inhibitor binds reversibly with enzyme to depress or nullify the catalytic action of enzyme. In competitive inhibition, the inhibitor will be structural analogue of substrate and competes with latter for bending groups of enzyme. The inhibitor bends with bending group of enzyme to form an enzyme-inhibitor complex. So the substrate affinity of enzyme is lowered progressively with rise in inhibitor concentration. This causes a progressive decline in velocity of enzyme catalyzed reaction.

A high concentration substrate may overcome competitive inhibition by forcing inhibitor out from its combination with enzyme. So a high substrate concentration produces the same maximal velocity Vmax as in absence of inhibitor. The higher the inhibitor concentration, higher is the substrate concentration necessary for attaining maximum velocity. The degree of inhibition depends therefore on the relative concentration of substrate and inhibitor and also on relative affinities of enzyme for the two.

Or

The degree of inhibition depends upon relative concentrations of substrate and inhibitor. *Eg:* An enzyme Succinic acid dehydrogenase catalyzes the conversion of Succinic acid to Fumaric acid. Many organic acids structurally related to succinic acid combine with enzyme thus inhibiting the reaction *Eg:* Malonic acid.

Many micro organisms like bacteria synthesize vitamin folic acid from para aminobenzoic acid (PABA). Sulfanilamide and other sulfa drugs are structural analogues of PABA. Hence sulfa drugs acts as enzyme inhibitor and occupy active site of some bacterial enzyme catalyzing this reaction.

Non Competative Inhibition

This is another type of reversible inhibition. But here inhibitor bears no structural similarity with substrate and binds reversibly with a site of enzyme other than its substrate binding site. So inhibitor may combine with both free enzyme and enzyme substrate complex. This probably brings about changes in three dimensional confirmation of enzyme to produce a failure of catalysis or of the release of the product. In this way the non competitive inhibitor lowers velocity of enzyme catalyzed reaction. It also reduces Vmax of the reaction. But non competitive inhibitor does not change the substrate affinity or km of enzyme, as it does not occupy the substrate bending site of latter. Moreover because the substrate and inhibitor bind with different sites of enzyme, the inhibition does not allow original Vmax to be regained by raising substrate concentration.

Eg: Various heavy metal cons (Ag^+, Hg^+, Pb^{2+}) inhibit the activity of Urease. Heavy metals form mercaptides with sulfhydryl (-SH) group of enzyme.

$$Enz - SH + Ag^+ \Rightarrow Enz - S\text{–}Ag + H^+$$

The established equilibrium inactivates enzymes that require a –SH group for activity. Because of the reversibility of mercaptide formation inhibition can be retrieved by removal of the heavy metal ion.

Uncompetitive Inhibition

An uncompetitive inhibitor bends at an allosteric site, but the bending takes place only with ES complex and not enzyme molecule.

Allosteric Regulation

Allosteric enzymes have one catalytic site where substrate binds and another separate allosteric site where the modifier binds. Activity of some enzyme is enhanced or inhibited by binding of specific low molecular weight regulatory metabolite with a specific site of enzyme called Allosteric site, which is different from its active site.

Such regulatory substances are called Allosteric effectors. They may be allosteric activators or inhibitors according as they enhance or reduce the enzyme activity by bending with Allosteric site.

The binding of an allosteric activator (positive modifier) to allosteric site may alter the configuration of substrate binding site to favor substrate binding. Allosteric enzymes are usually multiple subunit enzyme and the catalytic and allosteric sites may be located on different subunits. The bending of substrate to one submit of enzyme enhance substrate binding by other subunits. This effect is said to be positive co-operativity.

If the bending of substrate to one of subunit decreases the activity of substrate binding by other sites, the effect is said to be negative co-operativity. When an inhibitor binds to allosteric site, the configuration of catalytic site is modified such that substrate cannot bind there. In allosteric inhibition, the inhibition is not a substrate analogue. Allosteric inhibition is partly reversible when excess substrate is added. km is usually increased and V_{max} is reduced. *Eg:* Phosphohetokinase is allosterecally activated by Fructose-6-phosphate. The enzyme is allosterically inhibited by citrate and ATP; Aspartate transcarbamylase by tytidine triphosphate.

Michaeles Menten Equation (For single substrate Reaction)

$$V = \frac{Vmax\ (S]}{Km + (S)}$$

$$Km = (s)\left(\frac{Vmax}{V} - 1\right)$$

When $V = ½\ V_{max}$

$$km - (S)\left(\frac{Vmax}{\frac{1}{2}Vmax} - 1\right) = (S)$$

2.3. MECHANISM OF ENZYME ACTION

Catalysis is the prime function of enzymes. The nature of catalysis taking place in the biological system is similar to that of non biological catalysis. For any chemical reaction to occur, the reactants have to be in activated state or in a transition state.

Activation Energy

The energy required by the reactants to undergo the reaction is known as activation energy. The reactants when heated attain the activation energy. The catalysis reduces the activation energy and this causes the reaction to proceed at a lower temperature.

Active Site

As the substrate molecules are comparatively much smaller than the enzyme molecules, there should be some specific regions or sites on the enzyme for bending with the substrate such sites of attachment are variously called as Active sites or catalytic sites or substrate site.

Although the enzymes differ widely in their properties, the active site present in their molecule possesses some common features.

1. The active site occupies relatively a small portion of enzyme molecule.
2. Active site is a three dimensional entity. It is made up of groups that come from different part of linear amino acid sequence (or kept far apart from each other but brought close to each other)
3. The active site binds substrate molecules by relatively weak forces.

Enzyme Substrate Complex Formation

The prime requisite of enzyme catalysis is that the substrate (S) must combine with the enzyme (E) at active site to form enzyme substrate (ES) complex which ultimately results in product formation (P).

$$E + S = ES - E + P$$

A few theories have been put forth to explain mechanisms of Enzyme substrate complex formation.

Lock and Key Model or Fischers Template Theory

According to this model the structure or confirmation of enzyme is rigid. The substrate fits to the binding site (active site) just as a key fits into the proper lock. Thus the active site of an enzyme is a rigid and pre-shaped template where only one substrate can bind.

This model does not give any scope for flexible nature of enzymes; hence the model totally fails to explain many facts of enzymatic reactions. *Eg:* Effect of Allosteric modulators

Induced Fit Theory or Koshlands Model

Koshland in 1958, proposed a more acceptable model for enzyme–substrate complex formation. As per this model, the active site is not rigid and pre shaped. The interaction of substrate with the enzyme induces a fit or conformation change in the enzyme, resulting in the formation of a strong substrate binding site. Further due to induced fit, the appropriate amino acid of the enzyme are repositioned to form the active site and bring about catalysis. Koshland's model explains action of allosteric modulators.

Substrate Strain Theory

In this model the substrate is strained due to the induced confirmation change in the enzyme. It is also possible that when a substrate binds to preformed active site, the enzyme induces a strain to the substrate. The strained substrate leads to the formation of the product. The concept of substrate strain explains the role of enzyme in increasing the rate of reaction.

In fact a combination of induced fit model with substrate strain is considered to be operative in enzymatic action.

The formation of an enzyme substrate complex (ES) is very crucial for catalysis to occur. It is estimated 10^6–10^{12} times an enzyme catalysis reaction is faster than non catalysed reaction. The enhancement in rate of reaction is mainly due to

a) Acid – base catalysis
b) Substrate strain
c) Covalent Catalysis
d) Entropy effects.

Enzyme Characteristics

1. ***Colloidal nature*:** Enzyme molecules are grant on size; molecular weight range from 12000 to over 1 million. On account of their larger size, enzyme molecule poses extremely low rates of diffusion and form colloidal systems in water.
2. ***Catalytic nature*:** The enzymes act catalytically and accelerate rate reaction. They do not participate in these reaction and are recovered without undergoing any qualitative or quantitative changes. Thus, they in very small amounts catalyze transformation of large quantity of substrate.

   Turn over number: Measures catalytic power of enzyme which is defined as "The number of substrate molecules converted into product per unit time when the enzyme is fully saturated with substrate".
3. ***Specificity of enzyme action*:**
   a) *Absolute specificity*: Some enzymes are capable of acting on only one substrate. *Eg:* Urease acts only on Urea to produce ammonia and CO_2
   b) *Group specificity*: Some enzymes are capable of catalyzing reactions of structurally related groups of compounds. *Eg:* Lactate Dehydrogenase (LDH) catalyses interconversion of pyruvate and lactic acids.
   c) *Optical specificity*: Some enzymes will react with only one of the two optical isomers. *Eg:* Arginase acts only on L- argenine and not on its D-isomer.
   d) *Geometrical specificity*: Some enzymes exhibit specificity towards cis and trans forms. *Eg:* Fumarase does not react with malic acid which is an isomer of fumaric acid.
4. ***Thermostability*:** Being proteins they are very sensitive to heat
5. ***Reversibility*:** Enzymes are capable of bringing about reversion in chemical reaction. *Eg:* Lipase synthesise fat from glycerol and fatty acid and also hydrolyze them to this components.
6. ***pH sensitivity*:** Each enzyme acts best at a certain pH specific to its activity and slows down with any change.

Chapter 3

SUMMARY OF IMPORTANT PATHWAYS

3.1. ELECTRON TRANSPORT AND OXIDATIVE PHOSPHORYLATION

All enzymes catalyzed steps in oxidative degradation of carbohydrates, fats and amino acids in aerobic cells converge into electron transport chain and oxidative phosphorylation, the final stage in cell respiration.

An electron transport chain consists of a properly arranged and oriented set of electron carriers transporting electrons in a specific sequence from reduced Nicotinamide coenzymes and flavoprotein to molecular oxygen. The chain consist of a series of redox couples, at each step electrons flow from reductant of redox couple having a lower redox potential to oxidant of another redox couple possessing a higher redox potential. The electron chain located in plasma membrane or the mitochondrial respiratory chain in eukaryotes (located on the inner membrane) forms the final path for the flow of electrons from tissue substrates to molecular oxygen. The free energy liberated during the transmit of electrons along this chain is used in forming high energy bonds of ATP.

The NADH released from Glycolysis and TCA cycle directly enters into electron transport chain. The oxidation of NADH releases sufficient energy to drive the synthesis of several molecules sufficient enough to drive the synthesis of several molecules of ATP. However NADH oxidation and ATP synthesis do not occur in single step. Electrons are not transferred from NADH to oxygen directly; rather the electrons are transferred from NADH to oxygen along a chain of electron carriers collectively called the "Electron transport chain" or "respiratory chain".

The ETC consist of three large protein complexes embedded in plasma membrane called NADH Dehydrogenase, cytochrome bc complex and cytochrome oxidase. Electrons flow from NADH to oxygen through these three complexes.

Two small electron carriers are also needed to link these large complexes *i.e.*, Ubiquinone (Coenzyme Q) and Cytochrome C.

NADH to NADH dehydrogenase transfer reducing equivalence (H^+ and e) from NADH to CoQ. NADH dehydrogenase (Complex I) consist of at least 42 polypeptides. It binds to NADH and reoxidizes it to NAD^+ passing the two electrons from NADH to a prosthetic group called FMN. (Flavin Mono Nucleotide) to produce $FMNH_2$. 2 Iron sulfur protein molecule accepts 2 electrons from 1 $FMNH_2$ or $FADH_2$ molecule which simultaneously releases 2 protons into the medium. Each CoQ molecule accepts 2e from 2 Iron sulfur protein molecules and 2 protons from the medium to get reduced to QH_2. Each QH_2 in turn donates its electron through 2 Iron sulfur protein molecules to 2 cytochrome C molecule to reduce them and simultaneously releases 2 protons into the medium. A single electron is transferred from each cytochrome C_1 to C, C to a and a to a_3. Thus, one pair of electrons passes from an NADH or $FADH_2$ molecule through Cytochrome a_3 molecule to an Oxygen molecule, which will accept protons from the medium to form water.

NADH to NADH Dehydrogenase (Complex I)

NADH Dehydrogenase binds to NADH and reoxidizes it to NAD^+, passing two electrons from NADH to FMN to produce $FMNH_2$.

The electrons are then transferred to Iron Sulfur clusters (FeS). Complex II made of flavoprotein dependant succinate dehydrogenase complex carries reducing equivalence directly to CoQ from Succinate through its FAD and FeS clusters. Both the flavoprotein reductase transfer reducing equivalence from their respective substrates to Flavin prosthetic group (FAD, FMN). The reduced Flavin ($FADH_2$ and $FMNH_2$) transfers its electrons to Ferric ions (Fe^{3+}) of iron sulfur cluster reducing them to ferrous ions. Finally ferrous ions get re-oxidized to Fe^{3+} by flow of a electrons to CoQ which is consequently reduced to QH_2.

NADH Dehydrogemase to CoQ

Electrons from FeS cluster of NADH dehydrogenace are passed on to Ubiquenose (CoQ). This molecule can act as electron carrier by accepting 2e and $2H^+$ ions, where CoQ is converted to $CoQH_2$.

CoQ to Cytochrome bC_1 Complex

Compex III (Cytochrome reductase) transfers 2e from QH_2 to cytochrome C. Cytochrome bC_1 complex consist of Cytochrome b and Ccytochrome C_1 as well as FeS protein. Cytochrome act as electron carriers, as the electron is accepted the Iron atom of heme group changes from Ubiquinole through Cytochrome b, FeS and cytochrome C_1 components to cytochrome bC_1 complex and to next electron carrier Cyt c.

Since ubiquinol is two electron carrier where as cytochromes are one electron carrier, the pathway of electron transfer within cytochrome bC_1 complex is completed and involves $CoQH_2$ releasing first one e and H^+ ion to become $CoQH^+$ and then second e and H^+ iron to become CoQ (ubiquinone).

Cytochrome bC_1 Complex to Cytochrome C to Cytochrome Oxidase

Cytochrome C is peripheral membrane protein that is loosely bound to outer surface of mitochondrial membrane in eukaryotes.

It binds to Cyt bC_1 complex and accept an e *via* an Fe^{3+} to Fe^{2+} transition. Then it bends to cytochrome oxidase and donates the electron with Iron atom of heme to Cytochrome c and reverting to Fe^{3+} state.

Cytochrome Oxidase to Oxygen

Cytochrome oxidase (complex IV) contains cytochrome a and a_3. During electron transport iron atoms of cytochrome cycle between Fe^{3+} and Fe^{2+} states whilst the copper atoms cycle Cu^{2+} and Cu^+ (cytochrome a is paired with copper atom Cu_A and cytochrome a_3 is paired with different copper atom Cu_B. The Cytochrome oxidase reaction is complex, it transfers electrons from 4 cytochrome C molecules and $4H^+$ ions to molecular oxygen to form 2 molecules of water.

$$4\text{Cyt C (Fe } 2^+) + 4H^+ + O_2 \rightarrow 4 \text{ Cyt C} + 2H_2O$$

Oxidative Phosphorylation

Oxidative phosphorylation is the name given to synthesis of ATP that occurs when NADH and $FADH_2$ are oxidized by electron transport through respiratory chain (Unlike substrate level phosphorylation it does not involve phosphorylated chemical).

Whenever two electrons are transferred from one redox couple to another having a redox potential by 0.22 V that of former, the free energy change is sufficiently high to drive synthesis of high energy phosphate bond. There are such sites of ATP formation in ETC.

Oxidative phosphorylation is the formation of ATP by combining ADP and inorganic phosphate with the energy released by the transfer of electrons from different substrates to molecular O_2 through mitochondrial respiratory chain. Respiratory chain is responsible for large proportion of total ATP formation (when energy is trapped by oxidation of reducing equivalence such as NADH it is called oxidative phosphorylation).

Chemi Osmotic Theory

Peter Mitchell proposed Chemiosmotic theory to explain coupling of electron transport with oxidative phosphorylation. According to this, the free energy released during electron transport along the respiratory chain is used in translocating proteins actively from matrix to be cytosolic side of membrane in mitochondria. This creates an increased electrochemical gradient of protons across the inner membrane. The electrochemical potential of this proton gradient is utilized by ATP syntheses in phosphorylating ADP to ATP.

The electrochemical potential difference resulting from asymmetric distribution of H^+ ions is used to drive the mechanism responsible for formation of ATP. The inner membrane is impermeable to ions in general, but particularly to proteins which accumulate outside membrane creating an electrochemical potential difference across the membrane. This consists of chemical potential.

3.2. TCA CYCLE (CITRIC ACID CYCLE)

The Citric acid cycle is called Krebs cycle after Hans Kreb or Tricarboxylic acid cycle because of tricarboxylic acid intermediates. Citric acid cycle acts as a final common pathway for oxidation of

Carbohydrates, lipids and protein (because glucose, fatly acids and many aminoacids are all metabolized to Acetyl CoA or intermediates of the cycle). The cycle takes place in mitochondria.

Citric acid cycle is a series of reactions that brings about complete oxidation of Acetyl CoA to CO_2 and liberate hydrogen equivalents which ultimately forms water. The cycle comprises combination of a molecule of Acetyl CoA with 4 Carbon Dicarboxylic acid oxaloacetate to form 6 carbon tricarboxylic acid citrate. They follow a series of reactions in the cause of which 2 molecules of CO_2 are released and oxaloacetate is regenerated. Because of this, only small quantity of oxaloacetate is needed to facilitate the conversion of large quantity of Acetyl units to CO_2.

The citric acid cycle consist of a total of 8 successive reaction steps each catalyzed by enzymes.

The steps involved in The process are -

1. ***Condensation of Acetyl CoA with Oxaloacetate*:** Oxaloacetate (4C) and Acetyl CoA (2C) condense irreversibly to from Citrate (6C) catalyzed by Citrate synthase
2. ***Isomerization of Citrate to Isocitrate*:** Citrtae is converted to Isocitrate by an isomerization catalyzed by Aconitase. This is actually a two step reaction during which Cis Aconitase is formed as an intermediate.
3. ***Oxidative decarboxylation of Isocitrate*:** Isocitrate is oxidized to alpha keto glutaric acid and CO_2 by Isocitrate dehydrogenase. This enzyme requires NAD^+ which is reduced to NADH.
4. ***Oxidative decarboxylation of alpha keto glutarate*:** Alpha keto glutarate is oxidized to Succinyl CoA (4C) and CO_2 by alpha keto glutarate dehydrogenase complex. This is a complex of three enzymes and uses NAD^+ as co factor.
5. ***Conversion of Succinyl CoA to Succinate*:** Succinyl CoA is converted to Succinyl CoA synthetase. The reaction uses energy released by cleavage of Succinyl CoA bond to synthesize either GTP from Pi and GDP.
6. ***Dehydrogenation of Succinate to fumarate*:** Succinate is oxidized to fumarate (4C) by Succinate dehydrogenase. FAD is tightly bound to enzyme and is reduced to form $FADH_2$.
7. ***Hydration of Fumarate to Malate*:** Fumarate is converted to Malate by Fumarase. This requires addition of water.

8. ***Dehydrogenation of Malate to Oxaloacetate*:** Malate is oxidized to Oxaloacetate (4C) by Malate dehydrogenase. NAD^+ is again required by enzyme as co factor to accept the free pair of electrons and produce NADH.

The overall reaction may be written as :

$$CH_2CO - SCOA + 3NAD^+ + FAD + GDP + Pi + 2H_2O \text{ Krebs Cycle}$$

$$2CO_2 + COA - SH + 3NADH + FADH_2 + GTP + 2H^+$$

***Energy yield of Citric acid cycle (One turn of TCA Cycle)*:** The main purpose of TCA cycle is conversion of chemical energy into metabolic energy in the form of ATP. Hence the net yield of energy is 12 (As represented in Table 3) and the net energy yield of aerobic oxidation is 38 (As represented in Table 4).

Table 3: Energy yieled from TCA

Step	***Reaction***	***Formation of energy***	***ATP yield /mole***
3	Isocitrate – ketoglutarate	$NADH + H^+$	3
4	Keloglutarate – saucing COA	$NADHA + H^+$	3
5	Succuingl CoA – Saccinate	GTP	1
6	Succinate – Fumarate	$FADH_2$	2
8	Malate – Onaloacetate	$NADH + H^1$	3
		Total gain of ATP =	12

Table 4: Summary of ATP produced during Aerobic respiration of one glucose molecule

	Source	***ATP Yield***
1	Glycolysis Oxidation of glucose to pyreivate	2AH (Substrate level phosphorylation)
2	Production of $2NADH + H^+$	6 ATP Oxidative phosphyorylation and ETC
3	Formation of Acetyl CoA 2NADH + H	18 ATP ETC 4 ATP ETC 2 ATP (Substrate level phosphorylation)
		Total yield = 38 ATP

Formation of Acetyl CoA from Pyruvate

Under aerobic conditions Pyruvate is converted to Acetyl CoA by oxidative decarboxylation. The conversion of Pyruvate to Acetyl CoA is brought about by a Multi enzyme complex called Pyruvate Dehydrogenase complex. It is a mitochondrial enzyme. Pyruvate enters Mitochondria to be oxidatively decarboxylated to Acetyl CoA. This reaction requires TPP, Lipoamide, FAD, CoA and NAD as coenzymes. The pyruvate dehydrogenase complex has been made to be found up of Pyruvate decarboxylase component, Dihydro lipoyltransacetylase and Dihydrolipoyl dehydrogenase.

Since different enzymes activating are present as a single complex coordinated catalysis is possible. The oxidation of Pyruvate to Acetyl CoA is irreversible root from glycolysis to Citic acid cycle. Pyruvate dehydrogenase system generates reduced coenzyme (NADH). It also generates a high energy thio ester group in Acetyl CoA.

The oxidative decarboxylation is catalyzed by Pyruvate decarboxylase activity of Pyruvate dehydrogenase complex. Pyrvate is combined with TPP to form Hydroxyethyl TPP and CO_2 is liberated. The hydroxy ethyl TPP is oxidized to form an Acetyl group and then transferred to Lipoamide to form Acetyl Lipoamide. When oxidation of hydroxyethyl group occurs to acetÿl group the Lipoamide gets reduced to dehydrolipoamide. The acetyl group is then transferred to Coenzyme A (CoA SH) by enzyme Dihydrolipoyl transacetylase. The last step is oxidation of Dihydrolipoyl dehydrogene. The hydride is first transferred to FAD to form $FADH_2$ and then to NAD^+ to generate $NADH+H^+$. At the end of reaction three cofactors TPP, Lipoamedi and FAD are regenerated.

Mitochondria

Mitochondria are considered the power house of cells because energy released from oxidation of molecules is trapped as chemical energy in the form of ATP. The external membrane has an Assembly of enzymes and coenzymes which is not miscible with pool of enzymes and coenzymes in matrix. The inner membrane is highly selective in its permeability characteristics. All enzymes of TCA cycle are present in the matrix except Succenate dehydrogenase which is tightly bound to inner membrane. The final oxidation of carbohydrates takes place in mitochondrial matrix. The inner membrane convolutes into folds or crystae which contain numerous small spheres attached by stalks on inner surface. The spheres in inner mitochondrial membrane contain enzymes of election transport chain. The inner membrane also contains translocating systems that regulate the flow of ions and other substrates into and out of the matrix.

3.3. BIOSYNTHESIS OF FATTY ACIDS

C_2 fragments in the form of Acetyl CoA are fundamental building blocks in manufacture of fatty acid. Gibson and Wakil showed that actual molecule condensed in fat is not Acetyl CoA, but Malonyl CoA.

$$\text{Acetyl CoA} + CO_2 \rightarrow \text{Malonyl CoA}$$

Malonyl CoA is formed by condensation of Acetyl CoA with CO_2 catalysed by Acetyl CoA carboxylase. Malonyl CoA so formed condenses with a molecule of Acetyl CoA to form β oxoacyl enzyme complex and CoA.

$$\text{Acetyl CoA} + \text{Malonyl CoA} \rightarrow \beta \text{ oxoacyl} - \text{enzyme complex} + \text{CoA}$$

β oxoacyl undergoes reduction dehydration and loss of CO_2 to give CoA ester of C4 fatty acid in case of Butyryl CoA.

$$\text{Oxyacyl enzyme} \rightarrow \text{Butyryl CoA} + CO_2 + H_2O \text{ (C4 fatty Acid)}$$

Butyryl CoA can condense with a molecule of Malonyl CoA to form ester of Co fatty acid and so forth.

Denovo synthesis

Fatty acids are synthesized mainly by a *Denovo* synthetic pathway operating in cytoplasm. The major fatty acid synthesized *de novo* is palmitic acid a C_{16} saturated fatty acid. In addition there is a mitochondrial elongation system by which other fatty acids can be formed from palmitate in the body. The synthesis is with the help of multienzyme complex FA synthase carboxylase. The cofactor requirements include NADPH, ATP, Mg^{2+} biotin and bicarbonate. Acetyl CoA is the substrate and palmitate is the end product.

FA synthase complex is a dimer of two identical polypeptide monomers, each consisting 7 enzyme activities and acyl carrier protein.

Enzymes Involved in Substrate Binding and Condensation

Acetyl transacylase

Malonyl transacylase

Ketoacyl synthase

Reduction Unit

Hydratase

Enoyl reductase

Ketoacyl reductase

Releasing Unit

Thiokinase

The starting material for *Denovo* synthesis is Acetyl CoA (It is formed inside mitochondria from pyruvate. The inner membrane is not freely permeable to Acetyl CoA and carnitine transports long chain fatty acids. Hence Acetyl CoA units are delivered to cytoplasm as citrate which leaves mitochondria and is cleaved to oxaloacetate and Acetyl CoA in cytoplasm. The enzyme is ATP Citrate lyase. The oxaloacetate can return to mitochondria by malate shuttle or as pyruvate).

The first step is the synthesis is carboxylation of Acetyl CoA to form Malonyl CoA (The enzyme involved here is rate limiting enzyme and hence controls FA synthesis). It is catalyzed by Acetyl CoA carboxylase, a multienzyme.

Reaction Sequence

Acetyl transferase of domain 1 of multienzyme complex transfers acetyl group from Acetyl CoA molecule to Cys-SH of 3 ketoacyl synthase of same domain. Almost simultaneously Malonyl transacylase of the same domain transfers Malonyl group from Malonyl CoA molecule to adjacent pantothein –SH of acyl carrier protein of domain– 2 of other monomer. This forms an Acyl Malonyl enzyme.

3 ketoacyl synthesis next catalyses an attack by ketonyl carbon of cys bond acetyl group in methylene carbon of malonyl group held by pantathene SH. This releases CoA. The energy released during this stage is used to transfer acetyl to form 3 ketoacyl enzyme.

3 ketoacyl reductase transfers reducing equivalents from NADPH to 3 ketoacyl group reducing the latter to 3 hydroxyacyl group.

3 hydroxyacyl dehydratase removes water changes latter to 23 insaturated acyl enzyme.

The palmitoyl group formed at the end of even cycle is released as free palmitate from –SH of acyl carrier protein by hydrolytic action of esterase of synthase complex. The net reaction can be summarized as

Acetyl CoA + 7 malonyl CoA + 14 NADPH + $14H^+ \rightarrow$ 1 palmitate + 14 NADP + 7 CO_2 + 8 H_2O + 8 CoA

It constitutes transfer of malonyl group from one malonyl CoA molecule to acyl carrier protein to form a new acyl malonyl enzyme. Steps followed by this through a series of reactions namely condensation, reduction, dehydration and reduction are cycling repeated binding each time a new malonyl group resulting in the extension of acyl group by 2 C from malonate.

Seven cycles of FA synthase action produces a palmitoyl group. Each cycle utilizes one malonyl CoA and 2 NADPH molecule and releases one CO_2 and one water molecule. Acetyl group is used only once as a primer in the first cycle and as a source of C_{15} and C_{16} of palmitate. All the other C of palmitate comes from methylase C and CoA linked carboxyl C of each malonyl CoA molecule.

Odd number fatty may be formed by decarboxylation of Succinyl CoA, a C4 intermediate of TCA cycle.

Triglyceride Biosynthesis

Tryglyceride biosynthesis pathway is located in the mesosomal fraction of cell. Starting compound is glycerol phosphate which may arise from free glycerol and ATP (phosphorylation of glycerol) or from already phosphorylated C_3 compounds such as PGAL or PGA formed during photosynthesis. Glycerokinase catalyzes the reaction.

Glycero phosphate then combines with 2 CoA joined fatty acid units, resulting is phosphatidic acid which is their dephosphorylated to diglyceride.

Diglyceride is basic processor of both fats triglycerides and phospholipid (Lecithens). Diglyceride is acylated in presence of Acetyl CoA to give triglyceride.

Biosynthesis of Lecithin

Diglyceride combins with cytidine di phospho choline (CDP choline).

Oxidation of Fatty Acid

About 95% of fatty acids are present in esterified form; a small fraction is present as non esterified fatty acid or free fatty acid complexed to albumin.

- Triglycerides must be hydrolyzed to give free fatty acid and glycerol.
- These products are then oxidized. Glycerol can enter glycolytic pathway after conversion to Glycerldehyde 3 Phosphate.
- Fatty acids are broken down two Carbon at a time to release Acetyl CoA units.
- Acetyl CoA is fed into TCA cycle to furnish Hydrogen donors.
- Hydrogen donors fuel respiratory chain that produces ATP and H_2O.

β Oxidation

The principal method of oxidations of fatty acid is β oxidation. Other methods include v oxidation, w oxidation and peroxisomal fatty acid oxidation.

B oxidation theory was proposed by Fray Knoop

- According to this FA are oxidized by the removal of 2C atoms at a time. The C atom in the β position to carboxyl group is assumed to be attacked with the formation of the corresponding β ketoacid.
- The 2 terminal C atoms are split of as Acetyl CoA.
- A new carboxyl group is formed at the site of keto grouping so that a FA remains with 2C atoms less than the original.
- Again a new β carbon atom is attacked and 2 more C atoms are split of as Acetyl CoA.
- In this way FA are degraded by removal of 2C at a time until finally the stage of Acetoacetic acid is reached. Then it is split of to 2 Acetyl CoA units in case of even chain fatty acid.

Repeated β oxidation of odd chain FA release Acetyl CoA at each step and finally propionyl CoA mol.

Each sound of B oxidation involves 4 steps

1. Oxidation
2. Hydration

3. Oxidation
4. Thiolytic cleavage

Which results in splitting of 1 Acetyl unit and formation of fatty acyl CoA with 21 atoms less.

This would undergo the same series of reactions again until FA is completely oxidized to Acelyl CoA.

Enzymes of β Oxidation

β oxidation take place in mitochondria by several enzymes collectively called as FA oxidase system in mitochondrial matrix adjacent to respiratory chain which is found in inner membrane.

1. Activation of FA is done by Acyl CoA synthetase. Thiokinase in presence of ATP and CoA, converts FFA to active form.
2. ***Carrier Carnitines role*:** Acyl CoA (Activated FA) formed in cytosole should be transported to Mitochondria where β oxidation take place. For this it binds to Carnitine (β hydroxy v Trimethyl Anmnonium Butyrate).

   Thus, Acyl carnitines (long chain FA + Carnitine) is formed by enzyme Carnitine palmitoyl transferace I on innerside of outer mitochondria membrane.
3. Carnitine Acyl carnitine translocase transports Acyl carnitine.
4. Acyl carnitine then react with CoA catalysed by Carnitine Palmitoyl Transperane II attached to inside of inner membrane.

Chapter 4

GENETIC ENGINEERING

4.1. CLONING VECTORS

Stanley Cohen and Co-workers first reported the use of bacterial plasmids as molecular cloning vectors. The DNA that carries the desired gene to the host cell is called gene Cloning vector. It is also known as cloning vector, cloning vehicle or carrier DNA.

Genetic Engineering involves the transfer of a desired gene to a host cell that has to be manipulated. The desired gene may be hydrolyzed by cellular enzymes immediately after introduction. Further the chance for its expression is also very poor. In order to overcome these problems the desired gene is inserted into a suitable vector for gene cloning. Plasmid viral DNA's and Cosmids are used as gene cloning vectors.

- Plasmids are circular double stranded DNA's usually present in prokaryotic cells. They carry a foreign DNA of 5–15 kbp size to bacteria.

  Eg: pBR 322
- Some plasmids carry gene to plants

  Eg: Ti plasmids.
- Viral DNA's are linear or circular and single stranded or double stranded DNA's. Viral can carry DNA of 10–25 kbp size.

  Eg: Lambda phage and M13 phage DNA carry genes to bacteria

  Simian virus 40 DNA carry genes to animals cells.

- Cosmids are a type of constructed plasmid containing complementary single stranded sites (cos sites of DNA). They can carry DNA segment of 25–45 kbp size to bacteria.
  Eg: pHV 79

Without a vector, DNA molecule introduced into cell would be diluted out by cell division and eventually lost. Suitable cloning vector is chosen and cut with a restriction enzyme and the desired gene is inserted into a cleaved vector by using DNA Ligase. The resulting recombinant vector (rDNA) is then introduced to a host cell for gene manipulation.

There are a number of features common to all cloning vectors.

1. The vector must be able to replicate in the host cell. Thus it must have a replication origin and hence the vector can be amplified in the host cell.
2. Vector must contain a selectable marker for identification and isolation of bacteria containing vector. Marker in vector helps in identifying transformed bacterial colony. (Note:The efficiency of transfer motion is very low or only minority of cells in population take up and retain exogenous DNA)
3. Most commonly used vector contains a short sequence of DNA with many closely placed restriction enzymes cleavage site. It is known as MCS (Multiple Cloning site) and is helpful in gene insertion. Any two sites within an MCS can simultaneously cleave RE sites(that too without excising the vector sequence).

PLASMIDS

Plasmids are small, circular, double stranded, extra chromosomal DNA present in bacterial cells. They replicate independently due to the presence of an origin of replication.

Plasmids replicate along with bacterial chromosomes. The size of plasmid varies from less than 1 kbp to more than 200 kbp. Smaller plasmids are much desirable for gene cloning experiments.

The ideal cloning vectors must have the following characteristic features.

1. ***Size:*** The plasmid must be small in size. The small size is helpful for easy uptake of chimeric DNA by host cells and for isolation of plasmid without damage.

2. ***Copy markers*:** The plasmid must have one few genetic marker. Theses markers help for the selection of organism that has rDNA.
3. ***Origin of replication*:** The plasmids must have its own origin of replication and regulatory genes for self-replication.
4. ***Unique restriction sites*:** The plasmid must have unique restriction sites for common restriction enzymes in use.
5. ***Insertional inactivation*:** The plasmid must have unique sites for restriction enzymes in marker gene. This will help for the selection of recombinants by insertional inactivation method.
6. ***Pathogenicity*:** The plasmid should not have any pathogenic property.

Classification of Plasmids

On the basis of conjugative transfer, plasmids are classified into two categories. They are:

1. ***Conjugative plasmids*:** These plasmids are transferred from one bacterium to other during conjugation.

   Eg: F plasmid
2. ***Non-conjugative plasmids*:** These plasmids do not pass from one bacterium to another during conjugation.

   Eg: $ColE_1$ plasmids

On the basis of functions plasmids are classified into five types.

1. ***F–plasmid*** (Fertility plasmid) contains some genes expressing maleness in bacteria. *Eg:* F plasmid.
2. ***R–plasmids*** (Resistance plasmids) contain some genes giving resistance to bacteria against antibiotics and heavy metals etc. *Eg:* PSC 101 gives the organism tetracycline resistance.
3. ***Col plasmids*:** Col plasmids code for synthesis of bacterial toxin colicin. (Colicin kills other closely related strains of bacteria. *Eg:* Col E 1, Col B etc.
4. ***Degradative plasmids*:** Code for enzyme that degrade toxic substances such as toluene, xylene, parathion etc. *Eg: Psuedomonas putida* involved in breakdown of toluene.
5. ***Virulence plasmids*:** Virulence plasmids provide pathogenicity to the bacteria.

Natural and Based Plasmid Vectors

Some plasmids are isolated from bacteria and directly used for gene cloning without any modification. Such plasmids are known as Natural plasmid vectors. *Eg:* RP4 plasmid of *Pseudomonas*, Col E1 of *E. coli* and YEp and YIP of yeasts. Most of natural plasmids cannot be used for gene cloning. This is because:

1. They are large in size
2. They have no genetic markers
3. They have no unique site for common restriction enzymes in the marker gene
4. They confer pathogenicity to the host

In order to overcome these difficulties some unwanted portions are cutout from wild type plasmid and desired sequences are inserted into it. The effective cloning vectors created from wild type plasmids are called Based plasmid vectors, constructed plasmid vectors, derived vectors or artificial vectors. *Eg:* pBR 322, pSC101, pUC 8 etc. The constructed plasmids are of much use in gene transfer.

Amplification of Plasmid

The copy number of plasmid can be increased manifold. This process is called amplification of plasmid DNA. Chloramphenicol is added to medium (when cultures are in its exponential phase), which results in an arrest of chromosomal DNA replication and inhibit protein synthesis (and cell division). The cultures are then incubated for another 12 hrs for plasmids to replicate. This results in an increase in number of plasmids/cell. Sometimes the plasmid copy no: may reach to several thousand.

Plasmids can be categorized based on no of copies/cell as:

1. ***Relaxed plasmids*:** which are normally maintained at multiple copies per cell
2. ***Stringent plasmids*:** which have a limited number of copies per cell.

pBR 322

- pBR 322 is an artificial plasmid. This is derived from *E. coli* plasmid Col E1. 'P' indicates that it is plasmid. BR indicates the names of workers F. Boliver and Rodriguez who created the plasmid. 322 is the specific numbers to distinguish the plasmid from others. Features of the plasmid are as follows:

- pBR 322 is a circular, double stranded plasmid DNA. It consists of 4363 base pairs. This is regarded as the 'work house' of gene cloning laboratory.
- It has 20 restriction endonuclease sites.
- It has two selectable gene markers–tetracycline resistance gene (Telr) and Ampicillin resistance gene (Ampr).
- Tetr gene has unique sites for 6 RE. If a gene is inserted into any of these restriction sites, the tet gene becomes inactive.
- The Ampr gene has unique site for 3 RE. If a gene is inserted into any of these unique sites, the Ampr gene becomes inactive.
- Sequence other than Tetr gene and Ampr gene have unique sites for 11 RE. There is no insertional inactivation when a gene is inserted into any one of these sites.

Origin of pBR 322

Plasmid pBR 322 is constructed using three different naturally occurring plasmids. The Ampr gene was derived from RSF 2124. The Tetr gene was taken from psc 101. The origin of replication was obtained from pMB1 which is related to the plasmid Col E.

Advantages of pBR 322

1. pBR 322 is a small plasmid consisting of 4363 bp. (So it is resistant to shearing and increases the efficiency of uptake of bacteria during transformation).
2. The copy number of pBR 322 is 15. It can be increased up to 3000 by adding chloramphenicol.
3. pBR 322 is only 4.4 kbp in size. So it can carry relatively large DNA Fragments of 5–10 kbp (bacterial cells can uptake DNA's of 15 kb size)
4. pBR 322 has two selectable gene markers (Tetr and Ampr) for the selection of recombinants by insertional inactivation.
5. pBR 322 is used as a base plasmid for *in vitro* construction of derived plasmid vectors such as pUC 8, pU9, pUC 10 etc and cosmids. *Eg:* Somatostatin gene of man is introduced into *E. coli* through pBR 322.

Disadvantages

1. ***Instability*:** pBR 322 is lost in continuous culture (and in batch culture to a limited extent) This may create problems in large scale fermentation of recombinant.
2. ***Copy number*:** The multi copy (15–20 copies /chromosome) feature produce a gene dosage level too low or too high for a particular experimental condition.

   A low copy number is preferred when regulatory elements are being analyzed.

   High copy number overproduce the product of cloned gene. So pBR derivations with controllable copy number have been constructed.
3. ***Availability of cloning sites*:** Although it has 21 restriction sites, only 11 can be used. 3 fall within ori (This cannot be used for cloning purpose).

   Several pBR derivatives with multiple cloning sites have been constructed to enhance cloning versatility.
4. ***Host range limitation*:** Although it was originally constructed as an *E. coli* vector it replicate in a few instances in hosts like *Serratia marcescens*.

   (The host range is broadened by construction of plasmids called shuttle vectors).

pUC 18

The name is derived from the place of their initial preparation University of California by J Messing and J. Vieira. There vector have 2700bp. The features of the plasmid are as follows:

- It has an Ampicillin resistance gene
- It has an Origin of replication from pBR 322.
- It has a Lac Z' gene of *E. coli*
- Lac z' gene has a multiple cloning sequence.

  Lac z gene is a part of Lactose operon present in chromosomal DNA of *E. coli*. It is 3063 bp long and codes for the enzyme β galactosidase.

Of the 3063 bp, 123 bp at 3' end encodes for a short a helix polypeptide. Rest of the sequence code for long polypeptide of 980 amino acids (each long chain functions as a subunit of ß galactosidase. Four subunits held together by a helix form β galactosidase tetrameric enzyme).

The portion of Lac Z gene that codes for a helix is called Lac Z' gene. If Z gene is mutated the defective Lac Z gene produces long polypeptide but not a helix. Therefore β galctosidase has not been produced. Such *E. coli* cells are called Lac Z cells. Lac Z' gene is isolated from *E. coli* and inserted into vectors to use it as genetic marker. If plasmid possessing Lac Z' gene is introduced into Lac Z- cell α helix is produced by the plasmid gene and the cell produces active β galactosidase. This process is called a complementation.

β galactosidase reduces X-gal (5 bromo 4 chloro–3 indolyl-β-D galactopyranose) into a blue colored compound. If insertion of foreign gene inactivates Lac Z' gene, the plasmid fails to produce α-helix. So the cells remain white in X-gal agar medium. The white colonies represent transformed cells. The vector without insert sense blue colonies because there is no insertional inactivation.

Apart from this vectors of pUC family have a region that contains several unique restriction sites in a short stretch of DNA. This region is known as a polylinker on multiple cloning site (MCS).

Or

MCS is a synthetic DNA segment that has a cluster of unique sites for restriction enzymes. It is inserted into a vector with a view to increasing the number of gene cloning sites.

The size of MCS varies from 60–84 bp. MCS is usually inserted into Lac z' gene so that the insertion does not affect the function of Lac Z'. However cloning of a gene at any one restriction site of MCS inactive Lac Z' gene. This properly is useful for selection of transformants by a complementation.

Advantages

1. It is smaller than pBR 322. So can carry larger DNA segments (upto 12.4 kbp size)
2. The screening of recombinants is easier (than of pBR 322) and can be selected by growing in a medium containing X-gal.
3. MCS are used to increase the gene cloning sites.

4. Restriction enzyme of choice can be used to insert a gene into a vector.
5. As they have unique sites for many RL, DNA segments with different types of cut ends can be inserted into the vector.

Viruses as Cloning Vehicles

Bacteriophage serves as an excellent cloning vector. The cloning of single gene is usually best carried out using plasmids (since the insert be rarely larger than 2 kb). But for cloning of larger pieces of DNA (*Eg:* during gene library construction) these plasmids are not suitable, as larger inserts increases the plasmid size making transformation inefficient.

Bacteriophage has a linear DNA molecule, so a single break creates two fragments. Foreign DNA can be inserted between them and two fragments can be joined. Such phages when undergo lytic cycle in host will produce more chimeric DNA.

Wild type phage could accommodate only 2.5 kb of foreign DNA. Phage vectors are reconstructed by removing non essential genes and making vector DNA smaller so that larger insert can be accommodated in phage head during packing. *Eg:* γ phage, M B phage etc

Phage λ as Vector

Bacteriophage γ which infects *E. coli* are used as cloning vector. The process of infection by γ phage include,

1. ***Lytic phase*:** where phage particle injects its linear DNA into the cell, where it is ligated into a circle. It then replicates to form many phage particles which are released from the cell by lysis and cell death.
2. ***Lysogenic phase*:** where the DNA of phage integrate in to host genome by site specific recombination and remain there for a long period.

The phage consist of an icosahedral head and a flexible tail lacking contractive sheath. It has a linear duplex DNA with 48502 bp and molecular weight 32×10^6 daltons. On both ends of the duplex is cohesive single stranded extensions.

In duplex region more number of CG pairs are found at the left side and 47 pairs more at right side.

The cos ends allow the DNA to be circularized in the cell (each cos end is 12 kb long).

- Much of the central region of the genome is dispensable for lytic infection and is replaced by insert (upto 20 kb is inessential and up to 23 kb can be inserted)
- The recombinant DNA is then packaged within the viral particles *in vitro*.
- These viruses are then allowed to infect bacterial cells which have been plated out into agar (as a lawn of bacterial cells)
- Inside the bacterial cells, recombinant viral DNA is replicated.
- All genes needed for normal lytic cycle are still present in DNA and therefore multiplication of virus take place by cycles of cell lysis and infection of surrounding cells
- This gives rise to plaques on the background of lawn of bacterial cells.
- Cloned DNA can be recovered from the viruses in these plaques.

Gene Map of λ DNA

λ DNA is 48502 bp. It has 37 genes in definite clusters.

1. It has a cluster of 5 genes for head protein synthesis a cluster of 12 genes for tail synthesis and a gene for assembly of phages. All these 18 genes occur between 1 and 19400 bp.
2. There is a non-essential region between 19400 bp and 27975 bp. This sequence is useful in gene cloning.
3. Following the non-essential sequence there are 16 regulatory genes involved in recombination, integration DNA synthesis, regulation of DNA synthesis and lysis of host cell.

The λ phage is a temperate bacterial phage that infects *E. coli*. Although it has both lytic and lysogenic life cycle, the latter is more common in λ phage. The frequency of lytic life cycle is increased by treating the culture with mitomycin or exposing it to X-rays.

The λ DNA usually can carry DNA fragments and integrate into chromosomal DNA of the host cell. Therefore it is used as a vector of gene cloning in many instances. Infective properties on the other hand may also be taken as an advantage, particularly in the formation of DNA libraries.

The constructed λ DNA is of two types:

1. Insertion vectors
2. Replacement vectors

Insertion Vectors

When cloned into an insertion vector, the phage DNA is cleaved with a restriction enzyme that cuts it only once and the target is inserted into this site. No phage DNA is removed and therefore much lower size of DNA can be inserted. In commonly used vector λ gt 10, the Eco R 11 cloning site is in a gene which is deleterious to phage replication in certain host strains. This allows selection against non recombinant phage.

The restriction enzyme which has more than one site in nonessential sequence is used to cut the λ DNA. The DNA fragment found in between the cut ends of λ fragments are joined to form an insertion vector. The size of insertion vector depends on the size of non essential sequence deleted from the λ DNA.

Commonly used insertion vectors include λ charan 16, λ gt 10 etc. A foreign DNA of about 18 kb can be cloned in λ insertion vector.

λ gt 10–43 kb double stranded DNA and can clone fragments up to 7 kb.

> Non recombinant λ gt 10 gives cloudy plaques while recombinant λ gt 10 gives clear plaques.

Charon 16 -upto 9 kb foreign DNA can be cloned.

Replacement Vector

The replacement vector has two restriction sites for a restriction enzyme at non essential sequence. The DNA segment lying between the two restriction sites can be removed by using the restriction enzyme. And a foreign DNA can be incorporated at the site of removed λ DNA fragment. Since a foreign DNA replaces a portion of λ DNA, the vector is called a replacement vector. The replaced portion of λ DNA is known as stiffer region. *Eg:* EMBL3, DASH

Although here circular DNA or derivatives of it can be transformed into competent cells, the infective properties of phage are used in construction of DNA libraries.

- Replication of phage *in vivo* produces long linear molecules with multiple copies of genome.
- These concatomers are elevated at cos sites, to yield individual genomes which are packaged into phage particles.
- A mixture of phage coat proteins and DNA processing enzymes (a packaging extract) is used to package ligated DNA molecules into phage particles.
- The phage particles so produced is used to infect *E. coli* cells.
- The infection processes is very efficient and produce 10^9 recombinants per microgram of vector DNA.

The infected cells from packaging reaction are spread on an agar plate, which has been pre-spread with a high concentration of uninfected cells, which will grow to form a continuous lawn. Single infected cells results in clear areas or plaques within the lawn after incubation.

***Cloning DNA in bacteriophage*:** inessential DNA is cutout of DNA using RE and separated from essential DNA fragments by electrophoresis. Essential DNA and DNA to be cloned is legated and packaged in empty head and tails are added to gene infective virus.

Advantages

- The efficiency of gene transfer through a phage is high.
- Foreign DNA's up to 25 kbp can be packed in virus capsid and tranduced to *E. coli*
- Lytic life cycle is advantageous in constructing DNA libraries.

Limitations/Disadvantages

- Ligated DNA which is smaller or larger than 20 kb optimum, is too small a too large to be packaged.
- Sometimes many DNA join together by bare pairing between their cos sites and form relatively long DNA. Here packaging of DNA does not take place successfully.
- DNA has no easily selectable marker gene. So transformants should be identified by DNA hybridization method.
- The recombinant DNA may enter the lytic life cycle.
- The phage has narrow host range.

M 13

M 13 is a filamentous bacteriophage that infects *E. coli*. It consists of an outer protein coat called capsid and a DNA residing in the capsid. The DNA is circular and single stranded. It is made up of 6408 bp.

M 13 is a male specific and infects *E. coli* cells bearing F pilus. The life cycle comprises

- The phages attach to F pilus of *E. coli* anad injects its DNA.
- Inside *E. coli* cells, phage DNA stand (+) serves as a template for the sythesis of a complementary (–) strand.
- It becomes a double stranded DNA (ds DNA). Since it is a product of replication, it is called Replicative form (RF).

The RF does not integrate into the chromosomal DNA of *E. coli*, but multiplies repeatedly by rolling circle mode of replication (200–300 RFS their get accumulated in *E. coli* cells).

RF is converted into single stranded forms + and –, of the two strands + alone is packed into virus protein coat. The mature phages are extruded out of *E. coli* without breaking the cell wall.

M13 RF

The replicative form of M 13 phage is a circular double stranded DNA (6408 bp). RF has 10 genes essential for multiplication of M 13 phage. There is a short intergenic sequence (15) between origin of replication and gene number II. 15 is used for gene cloning.

Because the virus particles are very small relative to the number of the host bacteria, the host cells can be removed by low speed centrifugation. The virus particles can be collected from the supernatant suspension by high speed centrifugation.

M13 Based Vector

M 13 RF is well suited for gene cloning in *E. coli*. However, it has a unique site for only two restriction sites (Asa I and Ava II). Wild type phage is not an efficient vector and is modified to make it a good gene cloning vector.

J Messing *et al.*, constructed a series of gene cloning vectors from M13 RF. These vectors are called M 13 mp vector or M 13 based vectors.

Eg: M 13 mp 2, M 13 mp 7, M 13 mp 8 etc.

1. The M 13 RF (wild type) is isolated from *E. coli* cells infected with M13 bacteriophage.
2. The RF is cut with Ava I restriction enzyme at IS.
3. A Lac Z' gene isolated from Lac operon of *E. coli* is inserted into the cut site of RF using DNA ligase
4. The resulting circular DNA is called M 13 mp1.

   The Lac Z' gene encodes for a short polypeptide necessary for the formation of an active galactosidase from four subunits. When M 13 mp 1 is introduced into a Lac Z-mutant *E. coli*, active galactosidase is produced in *E. coli*. This enzyme converts colorless X-galactose in the medium into a blue substance. This help in blue-white screening.
5. The Lac Z^1 gene in M 13 mp 1 is subjected to site specific mutation to form M 13 mp 2.

   The Lac Z^1 gene has a GGATTC sequence near its 3^1 end. By mutation the second guanine (G) is altered to adenine (A) to form GAATTC in the Eco RI restriction site. The resulting vector is called M 13 mp 2.
6. M 13 mp 7, M 13 mp 8, M 13 mp 9, M 13 mp 10, M 13 mp 11, M 13 mp 18, M 13 mp 19 are constructed from M 13 mp 2. To construct these vectors M 13 mp 2 is cut with Eco R 1 at LacZ1 gene and a polylinker sequence is inserted into the cut site. The polylinker sequence is so designed that it does not affect the function of Lac Z^1 gene. All these vectors differ from each other in polylinkers sequence. *Eg:* M 13 Mp 7 vector.

It is a circular double stranded DNA. It is 7238 bp in size. It has a Lac Z^1 gene interrupted by a polylinker sequence at IS. Lac Z^1 gene serves as selectable gene marker for insertional inactivation. The polylinkers sequence is 66 bp in size. It has unique sites for 6 enzymes for gene cloning.

The M 13 mp 7, at polylinker has two restriction sites for Sal I and Eco R1. When it is cut with Sal I and Eco R1, a short segment between two sites comes out from the vector. The desired DNA replaces DNA segment during gene cloning. The recombinants are selected by a complementation.

Advantages of M 13 mp Vector

1. The M 13 RF exists in large number. So RI s' can easily be isolated and purified for gene cloning
2. As M 13 is a filamentous phage, relatively large DNA segments can be packed in phage particles (no strict size limit for DNA to be cloned)
3. As vector have MCS, DNA segments with different types of cut ends can be cloned.
4. All fragments of foreign DNA can be cloned at once by short gun cloning.
5. *E. coli* cells containing r DNA's can easily be selected by growing the cells in X–gal agar medium.
6. The host cell has not been lysed while releasing nature phages.
7. Single stranded DNA's with identical sequence can be isolated to DNA sequence.

M 13 RF – Plasmid Hybrid Vector

pEMBL8 is a derived vector constructed by combining pUC 8 plasmid and a part of M 13 RF. It is 3997 bp in size. The plasmid constitute Ampr gene, a MCS and Ori. The MCS is linked with LacZ' gene that serves as selectable marker.

The virus is allowed to infect *E. coli* cells with wild type M 13 phages. The latter serves as helps phages by proceeding various protein factors and capsids required for multiplication of recombinant phages.

Vector	*Marker*	*Size of Polyliner*	*Gene Cloning site*
M 13 mp 7	LacZ1	66	Eco R1, Bam H I, Sal I, Acc I, Hinc II and Pst I
M 13 mp 8	LacZ1	63	Eco RF, Sma I, Xma I, Bam H I, Hind III, Sal I
M 13 mp 9	Lac Z1	63	Eco RI, Hind III, Pst I, Sal I, Bam H, SMa I
M 13 mp 10	LacZ1	72	Eco RI, SSt I, Sma I, Xba I, Sal I, Ace I, Hind III, Pst I

Hybrid Vectors

Hybrid vectors will be having the characters of more than one kind of vectors, which makes it an efficient gene cloning vector.

Cosmid

Cosmid is an artificial plasmid containing cos-sites of DNA. It is formed by joining ends of linearised plasmid DNA with cos sites of DNA.

Thus it has two parts

1. ***Plasmid part*:** Which contain plasmid Ori, gene for drug resistance, unique RE sites etc
2. ***Viral DNA*:** A cos site of r phage which is essential for packaging of nucleic acid into protein coat.

   They lack structural and regulatory gene of DNA. Hence there is no lysis and integration of cosmid DNA in host cell. Instead the cosmid gets circularized and behaves like a plasmid.

   Eg: Col E 1 cosmid, p HC 79, pJB8, pWE etc.

   Cosmid was first constructed by Collins and Hohn in 1978.

Cloning Using a Cosmid Vector

1. A suitable cloning cosmid is selected
2. Cosmid is cut with restriction site I an appropriate RE. Thus cosmid is made into a linear form.
3. Fragment of foreign DNA is prepared using same RE
4. The foreign DNA is ligated with linear cosmid DNA
5. The foreign DNA will join with cosmids to form r DNA. Ligation occurs in such a way that concatamers of cosmids are formed.
6. Concatomers of recombinant cosmids will now be packaged *in vitro* into L phage heads.

   During packaging unit length are cut out at cos sites. In phage heads recombinant cosmid will remain linear.
7. These phages when infect bacteria, cosmids will be transferred into bacterial cells.
8. Inside the bacterial cell the recombinant cosmid becomes circular by base pairing at cos ends. Thereafter it replicates as a plasmid.

Salient Feature of Cosmids

1. Cosmid is a circular double stranded DNA.
2. It has two complimentary single stranded regions at both ends of DNA. This two cos ends form a duplex by base pairing.
3. At the cos site 3' end and 5' end of the same chain do not establish a covalent bond during circularization. So a definite nick is present in each DNA strand.
4. The nicks are retained in plasmid for number of generations.
5. The cosmid DNA does not code for phage proteins and host cell lysis. *Eg:* Cosmid pLER5

pLER 5 is a commonly used cloning vector suitable for cloning DNA fragments upto 45 kbp. It is 6 kbp in size. It is constructed from *E. coli* plasmid p BR 322 and two cos ends of DNA.

Ori, Tetr gene and MCS between Ori and cos site contribute the plasmid portion. The phage delivers DNA into *E. coli* after infection.

Advantages of Cosmids

1. Cosmids pickup relatively larger DNA fragments than plasmids do.
2. As they pickup large DNA fragments they are used to establish gene libraries of lower and higher organisms.
3. Some cosmids are constructed by joining a linearised plasmid DNA with DNA fragments of PI. bacteriophage that have cos ends. The PI bacterio phage has a genome of 115 kbp. So a DNA of 85 kbp can be packaged into head of PF phage. These cosmids help to clone large gene and gene clusters in bacteria.

Disadvantages of Cosmids

1. The packaged enzyme fails to pack recombinant cosmids into phage head, if any one of the two cos ends is missing.
2. Sometimes more than one recombinant cosmids join together to form a large DNA. If so packaging enzyme fails to pack DNA into phage head.

Yeast Artificial Chromosome (YAC)

Yeast Artificial Chromosomes (YAC) are desired cloning vectors used to clone large DNA fragments in yeast cells. They are linear in shape.

A YAC is made up of 3 essential regions

1. ***Autonomously Replicating Sequence (ARS)*:** Which is the replication origin of yeasts (It is highly unstable unless it is linked to a centromere)
2. ***A centromere*:** Is derived from yeast chromosome. It has 3 distinct regions I, II and III. The regeons I and III consist of 11–14 bp. They are separated from each other by region II consisting of 82–89 bp rich in TA. (Thus DNA of about 120 bp form CEN)
3. ***Two telomeres*:** On either side of YAC. They are rightly conserved and homologues to each other. Each telomere consists of multiple copies essential for the stability of YAC in yeasts.

The YAC having an ARS, CEN and two telomeres is called a mini chromosome. It behaves like an additional chromosome in yeast cell. *Eg:* YAC 2

For cloning purpose a selectable generic marker such as Amplicillin resistance gene (Ampr) is inserted between the telomere and ARS. The foreign DNA is inserted into YAC in between the CEN and telomere.

Advantages of YAC

1. YAC behaves as an additional chromosome in yeast cells. So yeast genome is not disturbed.
2. The reisolation of YAC from transformed yeasts and manipulation are easy.
3. Large DNA fragments irrespective of size can be inserted into YAC and introduced into yeast cells.
4. YAC can use all enzymes of yeast for its replication, transcription and translation. So it does not need special cases.
5. The selection of transformant yeasts is easy.
6. YAC is used to form genomic libraries of prokaryotes and eukaryotes.

Cloning Host *E. coli*

After ligation of target into vector the recombinant DNA is multiplied in the host. Various bacterial strains are available as host. Most common prokaryotic host is *E. coli*, which is also called as 'Workhouse of Genetic Engineering'.

E. coli is the most popular choice and genes from various sources including those from other bacteria, yeast, fungi, animals and plants, have been successfully engineered in *E. coli*.

Advantages

1. *E. coli* is simple to grow an large scale.
2. Higher biomass productivities can be obtained
3. First RE studied was from *E. coli*. So it becomes an obvious choice in the beginning.
4. Genetics and physiology of *E. coli* is amply known
5. *E. coli* can serve as a model molecular system for gene expression studies. Promoters gene regulators are well understood.
6. *E. coli* and its phages are most easy to manipulate
7. Suitable mutant strain of *E. coli* can provide an obvious host for gene bank made from another organism.
8. Even simple calcium chloride treatment allows transformation of *E. coli* with plasmid
9. Abundance of will characterized promoter- Lac.

4.2. TRANSFORMATION OF *E. COLI*

Cohen Ital (1972) showed that each treated *E. coli* cells are effective recipients of plasmid DNA. Almost any strain of *E. coli* can be transformed with plasmid DNA. It has been found that *E. coli* cells and plasmid DNA interact productively in an environment of calcium ions and low temp (0–5°C). For that a subsequent heat shock (37–45°C) is important, but not strictly required.

The method involves

1. Re suspended log phase cells in 50 m M $CaCl_2$ at about 10^{16} cells/m^1 and keeping them on ice for 30 min.
2. Plasmid DNA (0.1 mg) is then added to a small aliquot (0.2 ml) of these non competent cells.
3. Incubation on ice for further 30 min followed by a heat shock of 2 min at 42°C.
4. The cells are then transferred to nutrient medium and incubated for 30 min^{-1} hr.

Disadvantages of *E. coli*

1. *E. coli* do not usually get export of proteins into growth medium. Products retained within the cells.
2. Over expressed foreign protein form aggregates (inclusions) of denatured proteins. These are biologically non active and need to be solubilized and refolded to get active protein (however advantage is their resistance to proteolysis).
3. *E. coli* is not considered safe for r DNA technology because it occurs naturally in the gut of humans and risks are not predictable. Regulatory agencies are not in favor.
4. Endotoxin from *E. coli* often contaminates the proteins produced and is difficult to remove it.
5. *E. coli* cannot carry out post translation modification for genes from mammals.
6. Biological activity of the products by intra and extracellular proteins.
7. Damaged to products by intra and extracellular proteins.
8. Problems of plasmid stability, phage contamination also exist.

Other bacterial vectors used include *Bacillus subtilis*, *Streptomyces* etc. Yeasts are efficient host for eukaryotic studies.

DNA Isolation

Desired DNA can be obtained from cell DNA either by restriction digestion with restriction endonuclease or by mechanical shearing. From the mRNA the desired DNA can be made by using reverse transcription method. Desired DNA are also made from free nucleotides by chemical synthesis in a computerized machine called gene machine.

The desired gene to be cloned is residing in the cell DNA of source organism along with several genes. It should be isolated from other genes of cell DNA for gene manipulation. Two methods are followed to isolate a desired DNA from cell DNA of source organism.

Restriction Digestion

The total cell DNA from which the desired DNA has to be isolated from the source organism is isolated. The cell DNA is treated with a restriction enzyme. This enzyme cuts the cell DNA into a family of DNA fragments which differ in size and molecular weight. (cut ends of DNA may be blunt a cohesive depending upon the restriction enzyme).

In all restriction fragments, the desired gene is residing in any one of the DNA fragments. Therefore, all the DNA fragments are inserted into a suitable vector (because it is not known which one of the fragments possesses the novel gene.

If all restriction fragments are cloned at a time, it is difficult to screen the recombinants which received the desired gene. To avoid this problem, the mixture of DNA fragments is electrophoresed on an Agarose gel for a suitable time. During electrophoresis, DNA fragments get separated depending upon their sizes and form distinct bands. Each band is cut out of gel and DNA in the gel is isolated by washing it in wash solution. DNA fragments obtained from each of the bands are cloned in a vector at a time (This will make the screening easy).

Drawback

The restriction enzyme may cut the DNA at the middle of the desired gene and make it useless.

Mechanical Shearing

Cell DNA is subjected to mechanical forces to cut it into small fragments. Sonication with ultrasound cuts the DNA randomly and yields DNA fragments with the size of about 300 bp. The DNA fragments are blunt-ended.

Shearing forces can also be developed by passing the DNA solution rapidly through a hypodermic needle. Shearing force can also be given to DNA solution by stirring the solution in a blender. These shearing forces cut the DNA into fragments of 3000–30000 bp size. So the desired gene usually will not be affected.

Drawbacks

1. Every time mechanical shearing producers a new type of DNA fragment
2. The DNA fragments have an extra sequence in addition to a desired gene
3. This method is not suitable for eukaryotic DNA that has many introns.

4.3. ELECTROPHORESIS

The term electrophoresis describes the migration of charged particles under the influenze of an electric field. The important bimolecules such as amino acids, peptides, protein and nucleic acids etc process ionisable group and so at any given pH exist in a solution as electrically charged particle and will migrate to cathode or anode depending on the net charge.

Gel Electrophoresis

Gel Electrophoresis is one of the most basic and important tool in the study of generate engineering. It is used

1. To study the purity and intactness of DNA
2. To separate and identify DNA fragments
3. To analyze and characterize recombinant DNA molecule.
4. To determine the molecular weight of DNA fragments (by screening standard marker in parallel)

Advantages

1. Very simple and rapid
2. Resolve mixtures of DNA which cannot be separated by other methods.
3. As little as 0.05 mg of nucleic acid can be detected.

Under physiological conditions the phosphate group in sugar phosphate backbone is ionized in nucleic acid. Polynucleotide chāins of DNA and RNA are called Poly-anions and will migrate to positive electrode when placed in electric field. The factors affecting migration of DNA are:

1. Molecular weight of DNA
2. Shape and confirmation of the molecule
3. Magnitude of net charges in the molecule
4. Gel concentration
5. Applied voltage
6. Direction of electric field
7. Base composition and temp
8. Presence of intercalating dyes.

Materials and Equipments

Agarose or polyacrylamide is generally used for DNA separation by gel electrophoresis. The agarose gel have a lower resolving power and are used for fractionation of double stranded DNA larger than 200 bp.

Preparation of Polyacrylamide Gel Slab

PAGE is commonly used for sequencing of DNA fragments. It contains 30% acrylamide, 3% ammonium persulfate, 10 X TBE (Tris Borate buffer). These immediately after mixing is poured into gel apparatus on clean plates. There should be no trapping of air bubbles. Allow the acryl amide to polymerase up to 30 minutes in room temp. Then comb is inserted in get before polymerization. When polymerization is complete remove the comb.

Preparation of Agarose (0.7%) Gel Slab

The ingredients are agarose, distilled water, buffer and Ethidium bromide. Add agarose in distilled water and boil in dissolving the agar. Then cool it to 55°C. Add 10X TBE and Ethidium bromide. Pour the prepared gel in electrophoresis apparatus on clean glass plate. Insert a comb and after setting the comb is removed.

Buffers

All electrophoresis is carried out in an appropriate buffer, which is essential to maintain a constant ionization of molecule being separated. The buffer is chosen in such a way as to ensure efficient separation of molecule. *Eg:* TAE, TBE etc. (All there have pH range 7.5 to 7.8)

Electrophoretic Apparatus

Electrophoretic apparatus may be vertical or horizontal (Better resolution is obtained by vertical apparatus, but they are not easy to prepare and so horizontal are used widely because of ease of preparation, loading and better support). It is then connected to a Walter Schaffer power pack, which deliver power supply up to 200 mA.

Loading the Sample

Using disposable micropipette and gel loading buffer, introduce different samples in the wells formed in gel.

Connect electrophoretic apparatus to power supply with anode and cathode. The reservoir tanks of electrophoretic apparatus are filled with buffer.

Set the voltage to 30 V for overnight run or 80–100 v for 2–3 hours for agarose run. For PAGE pre run the gel for 30 min with 1V/cm. Then load the samples.

Detection

At the end of the run disconnect the cords used power supply. Remove the gel. Put the gel block in small tray. Pour the staining dye on the gel. Immerse the gel properly in staining solution. Stain for 1 hour with shaking intermittently. Remove the staining dye from tray and collect if for reuse. Carry out distaining of the gel. Give changes to water used till bands are distinctly visible. The DNA fragments can be seen as dark bands on the light blue background. The complete procedure takes 6–8 hrs.

The reagent used for revealing the portion of DNA band on agarose or poly acrylamide gel is fluorescent dye Ethidium bromide. This intercalates into DNA structure and can be seen by its orange/red fluorescence when excited by UV light.

(It is possible to quantify the amount of DNA of each band of gel by scanning densitometry. If DNA being separated is radioactive the bands can be detected by autoradiography of gel by southern blotting).

Recovery of DNA from Gel

The DNA can be recovered for any of the following purpose.

1. For restriction analysis
2. For preparation of probes
3. For ligation
4. For transformations

It can be recovered by 1) phenol extraction method 2) Freeze squeeze method 3) gel melting method 4) electro elution etc.

***Phenol Extraction Method*:** Gel slice is put in buffer solution for solubilization of DNA. Mixture is extracted with phenol. DNA remains in the upper aqueous layer. It is then ethanol precipitated. Alternatively

DNA from the gel may be sucked through Pasteur pipette – containing siliconised glass wool (removes agarose gel pieces) and subjected to phenol extraction and ethanol precipitation.

***Freeze Squeeze Method*:** This is suitable for normal agarose. It is reliable for small fragments of DNA (200–800 bp). Gel is put on a glass wool bed in small tube through which a hole is pierced. The tube is snap frozen in liquid Nitrogen and placed in larger tube and is centrifuged. As slice thaws aqueous phase is exuded and flows into larger tube while agarose is retained on glass wool.

***Gel Melting Method*:** Gel is melted at 70°C and diluted with low salt buffer cooled to 45°C and passed through DE 52 column. DNA is eluted through high salt buffer and then desalted through G25 column. (Reliable for 0.5 kb–10 kb).

Purification

For purification of DNA by chromatography following procedure can be used.

1. Use a small column. Add 0.6 ml DEAE Sephacel Slurry.
2. Wash with TE pH 7.6 containing 0.6 M NaCl - 3 ml or TE alone or TE containing 0.1 M NaCl.
3. Directly load DNA containing 0.1 M NaCl.
4. Wash column with 1.5 ml TE containing 0.3 M NaCl.
5. Elute DNA with 3, 0.5 ml washes of TE containing 0.6 M NaCl.
6. Extract the elute with phenol and chloroform. Recover DNA by ethanol precipitation.

4.4. BLOTTING TECHNIQUES

The phenomenon of complimentary base pairing between two DNA strands and between one DNA and one RNA strand having similar base sequence has been exploited for a very large number of sensitive techniques.

Some of the common techniques used are

1. Southern blotting or Southern hybridization
2. Northern blotting a Northern hybridization.
3. Western blotting

The name Southern blotting is derived from the name of inventor EM Southern. He developed a procedure for analysis of DNA and the methods extended for analysis of RNA and proteins subsequently have acquired the jargon terms Northern and Western blotting .

Southern Blotting

Southern blotting is the first blotting techniques developed by E.M. Southern who developed the procedure for transfer of DNA from gel to the nitrocellulose filter.

Steps involved:-

1. Separate the DNA from target cell and digest with restriction enzymes to get restriction fragments
2. The restriction fragments of DNA present in agarose gel after electrophoresis are denatured into single stranded form by alkali treatment.
3. They are then transferred into a nitrocellulose filter membrane; this is done by placing the gel on top of a buffer saturate filter paper, then laying the nitrocellulose filter membrane on top of this membrane. The buffer moves due to capillary action from the bottom filter paper to that on the top through the gel carrying with it denatured DNA present in gel. The DNA becomes trapped in the nitrocellulose membrane as the buffer passes through it. This process is known as blotting and takes several hours to complete.
4. The nitrocellulose membranes are then removed from blotting stack and DNA is permanently immobilized on membrane by baking it at 80°C.
5. Single stranded DNA has a high affinity for nitrocellulose filter membrane. The membrane is placed in a solution of radioactive single stranded DNA called probe.
6. DNA–DNA hybridization form between single stranded DNA and probe complimentarily.
7. After hybridization reactions, the membrane is washed to remove the unbound probes.
8. The membrane is now placed in close contact with an X-ray film and incubated for a desired period to allow images due to radioactive probes to be formed on the film. The film is then developed to several distinct bands corresponding to portions in

the gel of DNA fragments that are complimentary to radioactive probe used in the study.

Applications

1. Southern blotting is extremely sensitive and can be applied to mapping restriction sites around a single copy gene sequence in a complex genome such as that of man.
2. When mini satellite probe is used, the technique can also be used for forensic purpose for identification of minute amount of DNA.
3. The use of southern blot technique has also been done for analyzing the role of DNA methylation in gene expression.
4. Study of gene activation has also been benefited by this technique.
5. It is useful for the detection of mutated gene in genetic disorders.

Northern Blotting

Northern blotting is developed by Alwine, Kemp and Stark (1979) for detection of RNA in the sample. It is an extension of southern blotting. Initially Southern blotting could not be used directly to blot transfer RNA from gel to nitrocellulose membrane, because RNA did not bind to cellulose nitrate. Therefore a chemically reactive paper DBM (Diazobezyloxymethyl paper) is prepared from Whatsmann 540 papers. RNA binds covalently to this reactive paper and can be detected by hybridization with suitable radioactively labeled DNA probe.

Steps Involved

1. RNA is separated on the basis of size by electrophoresis through agarose gel.
2. The separated RNA are blotted into DBM paper
3. RNA after blotting are hybridized to radiolabelled DNA probe.
4. Autoradiography is carried out to locate RNA bands that are complementary to the probe.

Applications

1. To detect and qualitatively estimate hybridized mRNA.
2. Identify nuclear precursors of a particular mRNA
3. Estimate the molecular weight of hybridizing mRNA.

cDNA representing mRNA or cloned sequence corresponding to some particular part of the gene *Eg:* an Exon or an Intron may be used as a probe. When electrophoresis of RNA fractions is carried out and detection done above said probes, largest fragments corresponds to a size of gene and is probably a primary transcript. Smallest size is mRNA. In between bands show removal of some introns. Further information about introns can be gained from electron microscopy of individual nuclear RNA molecule. Each molecule is hybridized with DNA of intact gene and examined to see which introns have been removed. Any intron lost is shown as loop in DNA.

Western Blotting

In western blotting proteins are electrophoresed in polyarylamide gel, transferred into nitrocellulose or nylon membrane and protein bands are detected by specific interaction with Antibodies, lectins or some other compounds.

Steps Involved

1. Proteins are separated by polyacrylamide gel electrophoresis.
2. The protein bands are transferred into an intracellular or nylon membrane. This is achieved by capillary movement of buffer similar to southern blotting or electrophoresis (Electrophoretic blotting). Electrophoresis is more usually preferred, it is much faster and more efficient than capillary blotting (for both proteins and nucleic acid)

Protein bands obtained can be identified in a variety of ways.

1. Antibodies are commonly used as probes for detecting specific Antigens.
2. Lectins are used as probes for identification of glycoprotein. These probes may themselves be radioactive or a radioactive molecule may be tagged to them. Often the identification process is based on a 'sandwitch reaction'. In such an approach a specific Antibody is used to bind the antibodies bound to protein bands. These second molecule may be labeled with a radioactive enzyme or fluorescent dyes.

Proteins on membranes may be located by using radiolabelled specific Antibody to specific complimentary protein (Antigen Antibody). The Antibody is labeled with 1^{123}. Autoradiography is carried out and proteins can be located by staining with 0.1%. Comasie blue in 50%

methanol, 45% distilled water and 5% acetic acid and distained in same solvent.

ELISA technique may also be used for detection of specific protein (Antigen) bound on nitrocellulose.

Difference Between Southern and Northern Blotting

Southern	*Northern*
DNA is separated by electrophoresis	RNA is separated by electrophoresis
DNA has to be denatured before blotting	This step not required
Nitrocellulose membrane is used	DBM paper is used
DNA-DNA hybrid is formed	RNA-DNA hybrid is formed during blotting

4.5. METHODS OF DNA SEQUENCING

Once a gene or DNA fragment is cloned, its further study involves sequencing. For understanding the function of a gene it is essential to do characterization of DNA. This can be alone by knowing the nucleotide sequence. From the nucleotide sequence the amino acid sequence and molecular weight of protein can be understood. Comparison of the nucleotide sequence of newly deduced gene with known sequence of genes helps to correlate the various genes, sequencing and products of gene. Analysis of nucleotide sequence information is essential in molecular cloning studies. It also helps to know the regulation of gene expression.

There are four methods which are more commonly used for DNA sequencing.

1. Maxam and Gilberts chemical degradation method
2. Sangers dideoxyribonucleotide synthetic method.
3. Ligation and mediated PCR *i.e.*, LMPCR
4. Automatic sequencing

Maxam and Gilberts Method

This was the first readily available sequencing method. In this method, labeled DNA fragments are specifically cleaved by various chemical treatments. The cleavage is allowed to occur at G, G+A, C+T and T bases. These four reactions are sufficient to deduce the sequence of DNA fragments. Cleaved samples are electrophoresed and analyzed to read the sequence.

1. Label the 3^1 ends of DNA with ^{32}P.
2. Separate the two strands both labeled at 3^1 ends.
3. Divide the mixture into four samples each treated with different chemical reagent, having the property of destroying either only G T or A or C. The concentration of the reagent is so adjusted that 50% of target is destroyed.
4. Fragments of various sizes with ^{32}P at one end are produced.
5. Carry on electrophoresis on fragments in four samples in different lanes of gel.
6. Autoradiograph the gel.
7. Develop the sequence by reading the bands.

Sanger's Dideoxy Nucleotide Method

Fred Sanger is a twice Nobel prize winner. Sanger *et al.*, in 1977 developed a DNA sequencing method. This method is also referred to as plus minus method in which DNA polymerase was used to extend DNA chain. Subsequently he developed a more powerful method which is used single stranded DNA as template and $2^1 3^1$ dideoxynucleotides.

Normal natural deoxyribonucleotide has a 3^1 hydroxyl group which forms a phosphodiester bond. Dideoxynucleotide is a synthetic molecule and lacks 31 OH group and therefore cannot form phosphodiester bond during replication and on contrary will act as chain terminator. When such dideoxynucleotides are randomly incorporated in place of normal ones, it will result into fragments of various size. Dideoxynucleotides for different four bases are prepared and used in the reactions. These four samples are analyzed by Polyacrylamide Gel Electrophoresis. Electrophoresis is carried out at high temperature so as to get the DNA denatured. Long get is used and fragments differing even by one nucleotide are resolved the sequence can be directly read from autoradiogram.

1. Four reaction tubes are set up, each containing a single stranded DNA sample (Cloned in M13 phage) to be sequenced and all four dNTP's (radio actively labeled) enzyme DNA polymerase.

2. Each tube also contains either dideoxyribonucleotides of base A or C or D or G.
3. Fragments are generated as dideoxynucleotides will be randomly incorporated as chain terminators.
4. Fragments are separated in polyacrylamide gel by electrophoresis. Four sample are run in four lanes.
5. The gel is used for autoradiography to visualize bands.
6. DNA sequence can be deduced from the positions of bands

4.6. INTRODUCTION OF RECOMBINANT DNA INTO HOST CELLS

The rDNA inside the cell after introduction is safe and replicates independently of the chromosomal DNA. The desired gene present in rDNA express its characteristics in host cell. The cell containing rDNA is called transformed cell or recombinant cell or transformant. rDNA can be introduced into host cell by.

1. Transformation
2. Transduction
3. Transfection
4. Microinjection
5. Biolistics
6. Electroportation
7. Leposome Fusion.

Transformation

Bacteria intake the DNA fragments added to culture medium and express the characteristics of DNA's by a method called transformation. The ability of bacterial cells to take up DNA from medium is said to be competence which can be expressed by physical or chemical treatments.

To introduce rDNA into *E. coli*, rDNA is added to bacterial culture and treated with 50 mm $CaCl_2$ solution at Room temp. $CaCl_2$ adhers rDNA into surface of *E. coli* cells. The bacterial culture is then gently heated up to 42°C to induce cells to intake rDNAs.

Transduction

Introduction of fDNA into bacterial cells through a virus or bacteriophage is called transduction (rDNA of 38–50 kbp size can be introduced into bacteria using this method).

Phage desired vectors and cosmids have 2 cos sites one on either side of DNA. These cos ends help DNA to get packed inside the protein coat of phage to obtain recombinants.

The virus particles here are allowed to infect bacteria in culture. The virus injects its DNA into bacteria and rDNA get circularized inside the host cells.

Transfection

The direct intake of naked DNA from culture media by eukaryotic cells is called *transfection*.

Transfection in Plant Cell

Young plant cell can uptake DNA fragments added to nutrient solution by root system. But only few cells get DNA while most of the cells remain non recombinants. rDNA can also be introduced by using protoplast culture. PEG can be used to induce competence in protoplast to uptake DNA's having kamamycin resistance genes (Paszkovski *et al.* 1984). Protoplast taken rDNA can grow in medium containing kamamycin. However larger genes cannot be introduced in this method.

Transfection in Animal Cells

HeLA cells, Leukemia cells, liver cells etc can be treated with rDNA. $CaPO_4$ can be used to precipitate rDNAs on the surface of cells. The cells intake rDNAs by pinocytosis using cellular ATP's.

Electroporation

Introduction of rDNAs into cells by electric treatment is called *electroporation*. It is useful in obtaining rDNA in isolated plant, animal cells or bacteria. It is a rapid method to obtain more no of recombinants within short period of time.

rDNA and protoplasts suspension are mixed and kept together in small glass chamber. Two electrodes are placed in chamber and electric pulse of about 4–8 KV/cm is generated between two electrodes for 4–5 milliseconds. This electric pulse form transient pores in cell membrane and make it permeable to rDNA.

Limitations

1. Electroportion cannot be used to introduce rDNAs into intact plant cells.
2. Animal cells are sensitive to this method
3. If electric strength is too high animal cells may loose its viability.

Biolistics

Shooting of plant or animal cell by DNA coated tungsten particle for getting DNAs into cells is called biolistics suitable for plant, fungal and animal cells and cell organells like mitochondria and chloroplast.

In this method rDNA and tungsten particles of 1–4 diameter are used. Treatment with $CaCl_2$ or PEG precipitates rDNAs onto metal particles. rDNA coated with tungsten or gold are called microprojectiles. This are fired to plant or animal cells at a speed of 300–600 m/s using a microprojectile gun or short gun or particle gun.

Gun powder, Helium or electric power is used to provide propelling force to gun to drive micro properties into cells.

Microinjection

Microinjection refers to injection of rDNA's using micropipettes. Micropipettes are small capillary tube with open ends. Cells are transferred to microscope slide and rDNA's are sucked by micropipettes. Cells are held in position by sucking pipettes. Sharp end of micropipette are inserted by viewing through microscope and delivered to bacterial cells. Cells are transferred to fresh media to establish cell culture.

Limitations

Only few cells are made recombinants; technical skill and experienced workers required.

Agro Infection

Transfer of rDNA into plant cells by Agrobacterium is called Agroinfection.

Liposome Fusion

Liposome is small spherical vesicle made of phospholipids- Lipids like Phosphatedyl choline, cholesterol, phosphatidic acid, phosphatidyl resin etc are useful in making liposome.

In liposome, polar heads faces outward and non polar tails to centre. Liposome fuses with cell membrane and discharges its contents to cells.

Advantages

1. Any larger DNA's can be delivered to host
2. Protect fDNA from nuclease enzymes.

Selection of Recombinants

The host cell containing recombinant DNA is called recombinants or transformants. Inserted foreign gene express its characters in the recombinants. Recombinants occur along with non recombinants in culture; they should be selected and screened to have recombinants with desired properties.

4.7. SELECTION OF RECOMBINANT BACTERIA

Recombinant Bacterial forms can be identified by various methods like

1. Insertional inactivation of antibiotic resistance gene
2. Insertional inactivation of lac Z' gene
3. Radio Immuno assay
4. Colony hybridization
5. Nutritional complementation.

Insertional inactivation of marker genes and lac Z' genes confer antibiotic resistance and can be determined by replica plating method.

Colony Hybridization Method

Colony hybridization method was devised by M. Grunstein and D.S. Hognes in 1975. It is otherwise called *in situ* hybridization array. It is used to screen recombinants having desired gene from DNA library. It is rather screening method than selection process.

1. The transformed bacterial culture is diluted and poured into an agarised medium in petridish. The culture plate is incubated till bacterial colonies appear.
2. A nitro cellulose filter is placed over the colonies on agarised medium for some time. Some cells from each of the colonies get adsorbed in filter and form a replica of master plate.
3. The master plate is retained as such. The nitrocellulose filter is taken from petridish and placed in alkali solution. The alkali solution lyses bacterial cells and denatures DNA as well.
4. The filter is kept dipped in solution containing proteinase K, an enzyme that removes protein from filter. As a result DNA, alone gets bound to filter.
5. The DNA is fixed firmly on filter by drying at 80°C in between filter papers (This process is called baking).
6. The filter is placed in hybridization solution containing DNA probe or RNA probe labeled with P^{32}. The probe combines with target DNA in DNA print and forms DNA-DNA hybrid or RNA-DNA hybrid.
7. The filter is then washed in sterile saline citrate solution to remove unbound probes.
8. The Autoradiogram is matched with master plate. Black spots in autoradiogram locate exact position of recombinant colonies in master plate.
9. Recombinant colonies are picked up and cultured for future use.

Plaque Hybridization

This is a slight variant of colony hybridization devised by Benton and Davis in 1977 for screening recombinant phages. It is used for library screening by isolating recombinant phages. There

1. *E. coli* cells are cultured on agarised medium in petridish. The recombinant phage particles are allowed to infect *E. coli*. The

phage particles multiply in between and form plaques on bacterial colonies

2. A replica of plaques is made on nitrocellulose filter by placing filter ever plaques with original plate.
3. The filter is treated with alkaline solution to denature phage DNA.
4. The filter paper is baked at 80°C to fix DNA on it.
5. Filter is treated with hybridization solution containing P^{32} labeled DNA or RNA. As a result DNA-DNA or RNA-DNA hybrid is formed.
6. Filter is washed with standard saline citrate to remove unbound probes.
7. Autoradiogram is taken by placing X-ray film over filter. The probes form dark spots on X-ray film.
8. Colonies corresponding to dark spots are picked up from petridish.
9. Recombinant colonies are isolated from bacterial colonies for future use.

4.8. IMPORTANT MOLECULAR TECHNIQUES

RFLP

RFLP is Restriction Fragment Length Polymorphism; a laboratory technique used to compare DNA molecules of different species. A single Restriction enzyme produces fragments of different length from same stretch of DNA of different species. In this method

- DNA is isolated from several species
- DNA are cut with selected restriction enzyme
- DNA fragments are separated according to length
- The fragments are transferred from gel to nitrocellulose filter by Southern blotting.
- DNA fragments hybridizing with the probe can be visualized by autoradiography.

Applications

It is used to determine relationship between species. It shows genetic relationship between individuals. It is used in chromosome mapping. It is used to infer phylogenetic relationship.

DNA Fingerprinting

Distinguishing individuals according to their DNA prints pattern is called DNA fingerprinting. It is also called DNA profiling or molecular fingerprinting. It is used to assess whether an individual has genetic relationship with supposed individuals or not.

Satellite DNA is a non coding region of eukaryotic cells having tandem repeats of a sequence. It is also called mini satellite (rich in GC pairs, 9–40 bp in size). These satellite DNAs are unique to an individual. Satellite DNA's of different individuals are hybridized with single type of radio labeled probe DNA by Southern blotting. Autoradiogram is taken for DNA print pattern of individuals. From this DNA print pattern individuals.

Each and every mini satellite differs in length from corresponding mini satellite of unrelated individual due to variation in no of tandem repeats by deletion or addition of one or more sequences. It involves various stages

1. ***Isolation of cell DNA*:** Cellular DNA is obtained from blood or semen clotted on cloths, vaginal swabs taken from rape victims, hair or fresh cells or blood. DNA is isolated from the source material.
2. ***Restriction digestion*:** Cell DNA's are separated with RE that cuts on either side of mini satellite. Thus mini satellites get released from cell DNA.
3. ***Gel electrophoresis*:** Mini satellites are separated according to length by electrophoresis.
4. ***Southern blotting*:** Separated DNA fragments are transferred to nylon or nitrocellulose filter paper.
5. ***Selection of DNA probe*:** Alec Jeffry constructed DNA probes and are in use in USA. Dr. Lalji Singh (Centre for Cellular and Molecular Biology, Hyderabad) has developed DNA probe from mini satellite DNA for use in India.
6. ***Filter hybridization*:** Probe DNA binds with appropriate mini satellite and form duplex DNA.
7. ***Autoradiography*:** The radioactivity of probe make dark spots on X-ray film placed over filter (3hrs). An irregular ladder of dark spots develops on X-ray film for each DNA sample. Each ladder represents DNA print of an individual. All DNA prints on X-ray film together constitute DNA fingerprint pattern.

8. ***Analysis of DNA fingerprint pattern*:** All bands of DNA from source (hair, semen, vaginal swab etc.) should match perfectly with DNA print of accused. If match is 100%, individual is found to be responsible for victim.

Application

- To settle disputed parentage
- To settle murder cases
- To distinguish rapist
- Pedigree analysis
- Migration pattern (pattern of migration of ancient populations)
- Genetic analysis (agricultural crops, animal etc.).

Chapter 5

ENVIRONMENTAL BIOTECHNOLOGY

5.1. BIOLOGICAL INDICATORS

Biological indicators are species used to monitor health of an environment or ecosystem. They are biological species or group of species whose function, population or status can be used to determine ecosystem or environmental integrity. Such organisms are monitored for changes (Biochemical, Physiological or Behavioral) that may indicate a problem with their ecosystem. Biological indicators tell about cumulative effects of different pollutants in the ecosystem.

Pathogenic microorganism are likely to enter water periodically and most of them cannot survive in water for long periods (*Eg: Salmonella* and *V. cholerae*). Release of pathogens from patients or carriers will usually be small in members. Because of those reasons bacteriological analysis include search for certain groups of bacteria which are collectively referred to as fecal pollution indicators. There are three major groups.

1. The Coliform groups – typified by *E. coli*
2. The Fecal Streptococci groups- Typified by *Streptococcus fecalis*
3. The Suphite reducing Clostridia groups- Typified by *Clostridium perfringenes*

Coliforms

The term *Coliforms* refers to gram negative non spore forming rod shaped bacteria capable showing aerobic and facultative anaerobic growth in presence of bile salts and other surface active agents with growth inhibiting properties which are able to ferment lactose with the production of acid and gas within 48 hours at 37°C. They are Oxidase negative. The definition includes the genera *Escherichia*, *Citrobacter*, *Enetrobacter* and *Klebsiella*.

Coliform group is sometimes referred to as total Coliforms to differentiate them from Fecal coliforms. The major species in Fecal Coliform group is *E. coli* sometimes they are referred to as Thermotolerant. Determination of Fecal Coli forms provides a superior indicator of fecal pollution.

Fecal Streptococci

Certain Streptococci of fecal origin are classified as Fecal Streptococci. They belong to two main groups.

1. *Streptococcus fecalis, S faecism, S durans*- which are normally present in man and various animals.
2. *S.bovis, S equinus, S avium* – which do not usually occur in man

Apart from these species other serologically distinct streptococci occasionally present in feces are *S mitis* and *S salivarious*. These two species are inhabitants of mouth.

The occurrence of fecal streptococci are supplement to fecal Coli forms, indicates fecal pollution by warm blooded animals. Since Fecal Streptococci have a limited survival than outside the animal intestine and are not capable of multiplying in the environment their presence indicate a very recent pollution.

Morever certain species of *Streptococci* all associated with particular animals.

S. bovis – Sheep

S. equinus – Horse

S. *avium* – Poultry and other Birds

The comparison between the number of *E. coli* and other fecal streptococci in water can be used to determine the precise source of contamination.

Clostridium perfringens

C. perfrigens is the most important member of the group anaerobic sulphite reducing Closridia. It is normally present in human and animal feces, but in much fewer number than *E. coli* and Fecal *Streptococci.* Thus it is less sensitive as a direct indicator of fecal pollution. The resistant spores of *C. perfrigens* survive in water and in the environment for much longer time than *E. coli* and other fecal indicator organisms.

In the absence of other fecal pollution indicators the presence of *C. perfrigens* indicates remote intermediate fecal pollution.Therefore it is of much less importance in terms of immediate direct health risk. The presence of *C. perfringens* with *Coliform* organisms but not *E. coli* confirms fecal origin of pollution. These organisms are normal inhabitants of large intestine of humans and other warm blooded animals like cows, sheep's, pigs, poultry and are constantly released through their feces. Presence of these bacteria in water indicates fecal pollution, more specifically the recent pollution as these organisms cannot survive for longer periods in water.

The main sources of water borne pathogens causing enteric disease is water contaminated with feces of infected persons. Infected persons and carriers shed the pathogens through their feces along with other normal inhabitants of large intestine. Hence the presence of fecal pollution indicators can also be indirectly used as an indication for presence of enteric pathogens.

Among the indicators groups *Coliforms* are most important since they are ideal indicator organisms. *Coliform* groups in particular *E. coli* are constantly present in human intestine in large numbers and they are excreted in feces in large members each day.

Other Micro-organisms

Apart from fecal pollution indicators other micro organisms are also of considerable importance in water supplies. *Pseudomonas aeruginosa*, an important flourescent Pseudomonad may be derived from human or animal excrement, but its presence in feces is not universal. Because of this reason and its ability to multiply in water it is not suitable as an indicator of fecal pollution. Its presence in drinking water however is undesirable as it may affect colors, odors turbidity and taste of water.

Nuisance Organisms

Aquatic micro organisms which create problems of odor, colors and taste or which cause obstruction in flow are called nuisance organisms. These include

Slime forming bacteria – producing mucilaginous material favored by presence of organic materials *Eg: P fluorescens*

Iron bacteria – Transform soluble compounds of Iron to insoluble forms like Ferric hydroxide and get accumulated on its surface. The accumulation of insoluble material in metal piping system may affect the flow rate. *Eg: Sphaerotilus* (Deposits insoluble compounds as sheath). *Gallionella* (Secreted out to form stock)

***Sulfur bacteria*:** Capable of producing or tolerating extreme acidity.

Eg: Thiobacillus oxidises elemental Sulfur to Sulfuric acid. This produces acid PH which leads to corrosion of pipes.

Desufovibrio desulfuricans reduces Sulfur compounds like Sulfates to H_2S which produces considerable odor.

Algae

When water is exposed to sunlight algal growth occurs. The nuisance characteristics of algae are that they cause turbidity, discoloration, odor and taste in water. Some of them are known to produce toxins.

5.2. WATER BORNE DISEASES

Natural waters harbor diverse aquatic life which can range from microorganisms such as bacteria fungi and algae to higher organisms like fish. In unpolluted waters the density of these organisms remains less because of limited availability food and nutrients. On the other hand polluted water bodies especially Eutrophicated water bodies have a variety of microorganisms and usually harbor characteristic populations of flora and fauna. The water contaminated with sewage may contain various pathogenic micro organisms belonging to bacteria, viruses, protozoans in addition to the indigenous populations. The consumption of water or food contaminated with these microorganisms may result in Water borne diseases. The infected persons release large number of these pathogens which ultimately lead their way into municipal sewage. The pathogenic microorganisms are transmitted to the body of people through contact as by consumption of contaminated water and food (A representation of the important types of water borne diseases are represented in Table 5). Water borne diseases are classified different groups based on causal organisms. They are

1. Bacterial
2. Viral
3. Protozoan
4. Inset vectors
5. Algal

Table 5: Important types of water borne diseases

Organisms	***Diseases***
Salmonella	Typhoid
Salmonella paratyphi	Paratyphoid
Vibrio cholerae	Cholera
Escherichia coli	Intestinal diarrhoea
Leptospira	Leptospirosis
Clostridium bolulinum	Botulism
Other Salmonella and *Shigella spp Proteus spp*	Gasteroenteritis

Cholera (*Causative agent-Vibrio cholerae*)

Cholera affects millons in endemic areas and causes thousands of deaths especially during seasonal epidemics. Robert Koch discovered the organism and named it as *Vibro Cholerae* because of the characteristic shape. He originally referred to it as comma bacilli

Nearly 7–8 epidemics caused by *Vibro cholerae* affected different parts of the World. Last outbreak Occurred on Chennai in the year 1992 and other eastern parts of India.

V. cholerae is a gram negative bacteria measuring 0.3 μm in diameter and 1.3 μm in length. *V. cholera*e attaches itself to small intestine, grows luxuriantly in an alkaline environment in presence of bile salts (a condition prevailing in small intestine).

After successfully colonizing in intestine overcoming the peristaltic movement and other host defense mechanisms, it secretes large amount of Cholera toxin or Choleragen, which is responsible for diarrhea symptoms. Cholera toxin is a protein. The infection by cholera toxin finally leads to expulsion of sodium chloride from body along 10–12 litres of water in a day. This leads to dehydration and death.

In 1996 Mathew K. Waldor and John J Mekalanos of US reported stunning discovery about toxin. The toxin is a part of bacterium, but toxin released as virus get integrated into *Vibro cholerae* genome and now considered as a part of virus. Normally this virus remain silent within *V. cholerae* but during infection it gets activated. It then multiplies within bacterium and escapes to infect other *V. cholerae* (The phage is called CTX phage and CT stands for Cholera Toxin). Along with toxin, a specific pilus called Toxin Co regulated Pilus (TCP) is also produced on the surface of *V. cholerae*.

The released phages specifically attach to it and enter into bacterium. Vigorous viral multiplication results in the production of large amount of toxin causing severe diarrhea.

The toxin molecule of *V. cholerae* consist of 1 A and 5B subunits. The B subunit attach to ganglioside receptor on the surface of epithelial cells. The A subunit has A1 and A2 subunits. A1 fragment causes prolonged activation of cellular Adenylate cylate and accumulates cAMP in large amount leading to out pouring into small intestinal lumen large quantity of water and electrolytes leading to consequent watery diarrhea. The fluid contains more potassium and bicarbonate. This results in Cholera with massive water and electrolyte depletion.

The Cholera stool is typically colorless fluid with flecks of mucous said to resemble water in which rice has been washed; hence called rice water stool. The common complications are muscular cramps, renal failure, pulmonary edema etc. The clinical illness may begin with diarrhea and vomiting on 1–3 days as abruptly with massive diarrhea.

The treatment consists of adequate replacement of lost fluid and electrolytes (as ORS). Oral Tetracycline is also recommended.

Hepatitis

Viral hepatitis refers to primary infection of liver by hepatitis virus. The infective or Infectious hepatitis or Type A hepatitis occur sporadically or as epidemics affecting children and young adults and are transmitted through oral-fecal route. It is a sub acute disease of global distribution.

The large majority of infections are asymptomatic. The incubation period is 2–6 weeks the onset of disease may be acute or insidious with fever, malaise, anorexia, nausea and vomiting and liver tenderness. This usually subsides with the onset of jaundice. Recovery is slow over a period of 4–6 weeks. Very rarely rapid fatal fulminant hepatitis may occur. The disease is mild in children.

The causative agent of jaundice is Hepatitis A Virus (HAV) seen in feces of human. It is an enveloped RNA virus belonging to Picornavirus family. HAV is resistant to inactivation by heat.

Following ingestion, the acid resistance of HAV allows it to pass though the stomach into small intestine. The virus infects mucosal epithelial cells, replicates and spreads to adjacent cells *via* blood circulation and reaches liver. During the preicteric (prejaundice) stages of disease there is loss of appetite, fatigue, malaise, abdominal discomfort, fever etc. The virus is shed in the feces until the onset of jaundice or slightly longer. With the appearance of jaundice patient feels better, but liver remains tender and palpable. Jaundice persist for 1–2 or 3 weeks.

The epidemiology of type A Hepatitis resembles poliomyeletis. In developing countries infection is acquired in childhood. In India type A hepatitis is most common cause of acute hepatitis in children, but less frequent in adults. In affluent countries or developing countries with improved personal hygiene and sanitation, its incidence has been declining.

Diagnosis of type A hepatitis is based on.

1. Detection of Virions in patients feces by Immune Electron Microscopy (IEM), a technique by which Virions have reacted with specific antibody, becomes clearly visible from other particles under electron microscope. It is observed during pre jaundice phase and seldom in later phases.
2. Serological tests for viral Antigens
3. Demonstration of rising titer of Antibodies from patients. (I&M Anti HAV antibody appease during late incubation period)
4. Elisa kits for defector IGM and I&G antibodies are also available.

Improved sanitary practices and prevention of fecal contamination are ways to control the disease. A safe and effective formation inactivated, vaccine containing HAV grown in human diploid cell culture is available. A full course consists of two intramuscular injections of vaccine. Natural infection with HAV clinical or sub clinical leads to life long immunity.

Amoebic Dysentry

Protozoans cause intestinal diseases and are transmitted through contaminated water. The most important of water borne protozoan diseases is *Entamoeba histolytica* causing Amoebiasis, because of its great incidence. The most serious outbreak of amoebiasis in many years occurred in 1933 as an epidemic in Chicago during Century of Progress Exposition spread across nation and 1,400 cases and 4 deaths were reported.

People with amoebiasis have symptomatic cases of diseases with clinical indicators of diseases ranging from abdominal discomfort with slight diarrhea alternating with constipation to severe dysentery with blood and mucus in stools. Abscesses may be formed in liver or lungs or even in brain.

The pathogen is formed is tropical, subtropical and temperate regions. In human it invades mucosal and submucosal layers of colon. The toxin released by if dissolves mucous and submucous layers of colon leadings to bleeding. The disease at this stage is characterized by loose blood smeared stools- amoebic dysentery which is often fatal if untreated.

Nutrition of pathogen is holozoic. Fragments of epithelial cells and tissues, RBC's bacteria etc constitute the chief sources of nutrition. Life cycle of pathogen includes two stages.

1. ***A trophozoite stage*:** They are active non cysted forms that penetrate mucosa and submucosa, multiply and spread below mucosa to from a false shaped ulcer. Multiplication takes place by binary fission.
2. ***Encystment*:** The activity feeding trophozoite migrates to human of intestine where they undergo encystment. The cyst forms are spherical. Single nucleus in cyst divides twice at maturation resulting in four nuclei (Quadrinucleate stage). At this stage cyst comes out of body with fecal matter.

Through contaminated water it reaches new host and undergo excystation (alimentary canal) and become active trophozoite. Proper sanitary conditions in drinking water and disposal of fecal matter are essential to control the disease. Treatment of drinking water including disinfection, in water distribution systems is only control measures. For household purposes boiling water is an effective method. Preventing the contamination of water sources is the most effective control. In severe cases fever chills accompany dysentery. Laboratory diagnosis depends on identification of *E histolytica* in stools.

5.3. DETERMINATION OF SANITARY QUALITY OF WATER

Drinking water, polluted water from natural source and treated effluent have to the examined regularly for determining their quality. Bacteriological examination of water actually employs three different techniques. Different samples are selected and sanitary quality will be assessed based on the samples collected (Table 7).

1. Total count
2. Membrane filter method
3. ***Multiple tube method***: The organisms of interest in bacteriological examination are indicator organisms Coliforms, fecal coliforms and fecal Streptococci. Each technique has its own application and its own advantages and disadvantages.

Standard Plate Count (SPC)

The total count or colony count or Standard plate Count (SPC) gives total number of micro organisms comprising bacteria yeasts and moulds in water sample. It is usually done by pour plate method. In this method a known volume of water sample is mixed with molten media (Yeast Extract Agar or PDA or NA) and allowed to set. One set of plates are incubated at 37°C for 24 hours and other set at 20–20°C for 2–3 days. Most bacteria capable of growing in water will grow better at 22°C than at higher temperature. Organisms which grow at 37°C will grove less readily in water. Since these two groups of organisms differ in their significance they have to be counted separately.

Though the colony counts are not essential for assessing the safety of potable water supplies they are useful for indicating the efficiency of certain processes in water treatment such as coagulation, flocculation and disinfection. In addition they also indicate cleanliness of distribution system. They can also be used to determine suitability of water supply for large scale preparation of food and drink.

Membrane Filter Method

When numbers of indicator organisms in water is very low direct isolation on solid media is not practicable and other methods must be used by which large volumes can be examined. Membrane filter techniques are widely used for enumeration of bacteria from water sample. In this method bacterial cells are filtered through a membrane and placed on a suitable solid medium. On incubation these cells produce visible colonies which can be counted.

Filter apparatus

The membrane filter apparatus consist of a base supporting a porous disc. There is a filter funnel which can be fitted with base by means of screw threads or clamps. The filtration unit is usually connected to vacuum pump. The filtration apparatus has to be sterilized before use. Membrane filters are made up of cellulose acetate esters or cellulose citrate esters. They are 47 µm in diameter and have a pore size of 0.45 µm. Membranes usually have black grid marks to facilitate counting. Sterilized filters are available or they can be sterilized by autoclaving at 110°C for 10 min. Absorbent pads of the size same as filters and 1mm in thickness which are soaked in sterile nutrient medium can be used to provide nutrients for the growth of cells.

Filtration procedure

A sterile absorbent pad is placed at the bottom half of sterile petridish using sterile forceps. A suitable broth is poured over the pad so that if is completely saturated with the broth. The petridish is closed and kept aside. Touching only the other side of the holder the sterile filter paper holder assembly is replaced tightly without destroying the membrane. The sample is poured into funnel wetting the filter. For smaller sample size 20–30 ml of sterile buffer can be proud first to give an even distribution of sample. The sample is sucked through filter using vacuum pump. The filter is rinsed with 20–30 ml of sterile buffer and sucked through filter. The filtration apparatus is later disassembled using sterile forceps and membrane filter is transferred into nutrient medium in petridish with the grid side up. The Petridishes are incubated for given period of time and the colonies formed on the filter are counted.

Advantages

1. Speed of sampling

2. Results obtained within 22–26 hours
3. Larger volume of sample with low bacterial content can be analyzed.
4. Since direct count is made results are precise
5. Instrument is potable, hence useful for field studies.

Limitations

1. Samples containing large quantities of colloidal materials or suspended solids such as iron manganese, flocs, clay or algae can clog the pores and prevent filtration
2. If bacterial counts are high, in such cases a higher sample dilution must be used to minimize the effect of turbidity.
3. Industrial waste water may contain zinc copper or other heavy metallic compounds which adsorb into membrane surface and interfere with subsequent bacterial development.

Multiple Tube Test (Most Probable Number Method or MPN)

This test is based on statistical values of results obtained. This method detects and estimates various groups of micro organisms in water samples. This method actually involves three stages: Presumptive test, Confirmative test and Completed test. In this method, measured volumes of water as one or more dilutions are added to a serious of tubes containing suitable liquid differential medium. It is assumed that on incubator of each tube which received one or more test organisms in inoculums will show growth and characteristic change produced by organism. Provided that negative results occur in some of the tubes, most probable member of organisms in 100 ml of sample can be estimated from number and distribution of tubes showing positive reaction. To confirm that positive reactions presumptive tube is due to growth of specific indicator organisms, they are sub cultured on confirmatory media. Positive reactions in confirmatory test confirm presence of indicator organisms. Completed test involves biomedical test for identification and confirmed counts are calculated by referring the probability tables.

Procedure

MPN can be done either in three dilutions or five dilutions of sample. The dilutions used must be in consecutive order. *Eg:* 10 ml, 1 ml and 0.1 ml or 1 ml, 0.1 ml and 0.01 ml and so on for three dilutions and

100 ml, 10 ml, 1 ml, 0.1 ml and 0.01 ml or 10 ml, 1 ml, 0.01 ml and 0.001 ml for five dilution technique. In presumptive test, samples are inoculated each in 10 ml of corresponding both. If sample volume is 10 ml or more then the medium used should be of double strength. After 24–48 hours the results are noted based on acid production and/or gas production in tubes.

Advantages

1. Interpretation or results requires minimal experience and training as results can be got by observing for presence of gas or no gas and acid or no acid.
2. Water samples with high turbidly can be analyzed since there is no apparent deleterious effect.
3. Because of dilution used in range of 1:0 or 1:00 toxic substances present in sample can be diluted out.

Limitations

1. MPN procedure takes a very long term as for confirmed test result
2. MPN results are probability calculations and cannot be accurate
3. Requires more glassware and chemicals
4. False positive results are common occurrence

***Comparison of Membrane Filter and MPN technique*:** A detailed account of the difference between membrane filtration and MPN technique is given in Table 6.

Table 6: Comparison of Membrane filter and MPN techniques

Membrane filter	*MPN*
1 Comparatively smaller	Larger quantities of culture media and glass wearer required. Sterilization need larger autoclave
2 Membrane Filter is expensive	Uses less expensive materials
3 More accurate procedure	Statistically based method and possibility of errors as they may clog the membrane
4 Not suitable for turbid samples	Useful in analyzing turbid water
5 More convenient	Less convenient as large quantities of materials required
6 Incubator requires 24 hrs	Incubation requires 48–72 hrs
7 Suitable for field tests	Not suitable for field tests

Table 7: Selection of Methods for different sample

	Samples	*Parameters*	*Method*
1	Shellfish harvesting Water	Total coliform Fecal coliform	MPNMPN
2	Marine and estuarine waters	Total coliform Fecal coliform	MF/MPNMF/MPN
3	Treated industrial water (non-Chlorinated non toxic and low solid waste)	Fecal coliform	MF
4	Toxic industrial waste (Metal phenols) and high solid waste	Fecal coliform	MPN or Alternate procedure
5	Municipal and industrial effluent	Total coliform Fecal coliform	MPNMPN
6	Chlorinated secondary efficient	Total coliform Fecal coliform	Two step MF MPN or Alternate procedure

5.4. MUNICIPAL WATER TREATMENT

Physical Methods of Treatment

Water entering the treatment plant may contain large floating and suspended solids like leaves twigs and silt particles. Removal of such particles is the preliminary step in treatment process.

Screening and straining

This is the first stage preliminary treatment process. Coarsc screens with openings of 75 mm are used to prevent large objects. Main screens are provided in the form of mesh with openings of 5–20 mm through which the incoming water passes. To prevent overloading of mesh with collected solids, the mesh is slowly rotated so that solids can be removed at regular intervals.

Micro straining

Micro strainer is a finely woven stainless steel mesh with apertures of size about 20–60 μm. These micro strainers are effective in the removal of relatively small solid particles and algae from water. Frequently clogging of mesh is a problem in micro strainers because of the small aperture. So mesh is usually designed as of drum type which can be rotated around a central axis and mesh is continuously washed and cleaned by high pressure sprays.

Aeration

Aeration is a water treatment process in which water is brought into intimate contact with air. This is to increase oxygen content, reduce CO_2 and remove methane, hydrogen sulfide and other volatile organic compounds responsible for taste and order.

For drinking water treatment aeration can be achieved in many ways. Two important ways are water fall aeration or bubble aerators. In fist type water is dispersed through air in thin sheets or fine droplets. In the second type water is mixed with deposited air.

Sedimentation

Raw water contains various suspended matters such as reduced Iron and Manganese, bacteria and organic colloids as impurities. This can be allowed to settle, and removed when water stands still. Sedimentation is achieved when water is stored in reservoirs itself. Sometimes as in case of clarification of river water, sedimentation can be used without pretreatment and is called plain sedimentation (In other cases pretreatment like aeration, coagulation and flocculation are required before sedimentation.

Flocculants are added after which flocculated materials settle out in sedimentation tanks in the form of sludge. Most of micro organisms are also removed in sedimentation.

Coagulation and Flocculation

Coagulation and flocculation make finely divided suspended and colloidal matter in water to agglomerate and form flocs. This enables removal by sedimentation or filtration. Substances that cause turbidity and color are usually removed by flocculation and coagulation. These problems are mainly associated with surface waters. Sometimes algal growth and animal debris carried by surface run off also causes of turbidity.

Flocculation involves gentle mechanical agitation which causes formation of large flocs. Coagulation uses certain chemicals called coagulating agent or coagulants which enable flocculation. The most commonly used coagulant is Aluminium sulfate commonly called Alum. In normal conditions aluminum is converted to an insoluble form and thus removed from water.

Coagulation and flocculation are usually followed by sedimentation in large reservoirs for holding period.

Sand Filtration

Filtration of water is an important stage in treatment of potable water to achieve final clarification. Usually it follows sedimentation with or without coagulation. Filtration is a process in which water is purified by passing through a porous material or medium. Removal of suspended matter in porous media bed depends upon mechanisms such as gravity diffusion, sedimentation and hydrodynamic processes. Once transported into pores of bed the suspended matter is held thereby attachment mechanisms due to physico- chemical and intermolecular forces. Thus a bed of porous media can remove particles smaller than usual pore size.

A bed of fine sand can act as a medium for filtration and these filters are called sand filters. Sand filters are highly efficient in water treatment and cause removal of 99% micro organisms form raw water.

Sand filters are advantageous over other treatment due to various reasons

1. The quality of water produced is good
2. Micro organisms are removed
3. No chemical agents used (no unpleasant taste due to that)
4. Cost of construction is comparatively low
5. Simple in design and operation

These are two types of sand filter – Rapid and Slow sand filters.

Rapid Sand Filters

Rapid sand filters are usually built open with water passing down filter by gravity. It has a filtration rate of 5–15 $m^3/m^2/$ hr. The tank is made of either steel or concrete with depth of 3.0–3.5 m. Coarse sand (0.445–0.75 m thick) is used in filter medium. This is supported by coarse gravel of 0.60 m thickness. There is an inlet channel at one end through which water filtered enter the tank. An inlet valve controls inward flow of raw water. In lowest layer of gravel there is collector system of perforated pipes or ducts which are connected to main collecting channel. Filtered water is collected by perforated pipes and sent to storage tank through main collecting channel. There is an outlet controller which maintains the discharge flow.

In rapid sand filter removal of impurities from raw water is affected by different processes like straining, sedimentation, adsorption and bacterial and biochemical processes.

Types of rapid filters

1. ***Pressure filters*:** High pressure applied to filter bed brings out filtration process. Drawback in this filtration is installation operation and maintenance (not easy)
2. ***Upflow filters*:** Employ coarse to fine filtration. Sand alone is used as single filter medium. Bottom layer is made of coarse sand which filters out most of impurities. Upper layer contains fine sand which filters out remaining impurities.
3. ***Multiple media filter*:** Filter bed is composed of several different materials which are placed on coarse to fine in direction of flow.

A simple type of multimedia filter, the dual filter consist of a sand layers (0.3–0.5 m thickness) with 0.4–0.7 mm effective size bottom layers above which is a layer of crushed coconut husks of 0.5–0.7 m thickness with in effective size of 1.0–1.6 mm.

Slow Sand Filtration

A typical slow sand filter consists of 3 m deep tank which is open at the top. The area of tank varies with amount of water to be treated and usually is four times to several hundred square metres. An under drain system called filter bottom is placed at the bottom of tank though a main collecting channel. Filter bed has total thickness of about 1–1.2 m above which water to be treated stands to a height of about 1–1.5 m. The stone sand filter with number of influent and effluent lines are fitted with valves and control devices to maintain level of raw water and filtration rate constant.

Floating algae and other wastes may accumulate on water surface as a scum during filtration. To remove this, outlets can be provided at corners of the tank.

During operation raw water tank enters tank through inlet. Infiltration takes place when water percolates down filter bed. The filtered water is collected by under drain system.

Mechanism of Sand Filtration

Impurities from raw water are removed by a combination of different processes such as straining, sedimentation, adsorption and biochemical and micro biological processes. The removal of organic compounds depends upon the type of filtration employed. The major difference between Rapid and Slow sand filter is expressed in Table 8.

Table 8: Difference between Slow sand filter and Rapid sand Filter

	Slow sand filter	*Rapid sand filter*
1	Detention of pure of water is high. Low filtration rate	Less detention time. High filtration rate
2	Microbial activity maximum	Filtration does not involve biological layers
3	More space	Less space
4	Purify 5 million gallons of water day	200 Million gallons of water a day
5	Sand layer scraped out after certain days of operator	Back washing has to be done periodically once in 12 hrs

As filtration proceeds a gelatinous film accumulates around each grain of sand, composed of millions of bacteria, protozoa, filamentous algae and other microorganisms which is called schmutzdecke. This layer is usually formed in upper three or four inches of sand. This slowly closes up the pores between sand grains thus making filtration more effective.

Straining removes suspended particles that are too large to pass through pores of filter bed. Straining take place at the surface of filter. Periodically the accumulated impurities are removed off by scraping top layers.

Sedimentation removes fine suspended solids when they are deposited into surface of sand filters. The larges surface area of fine parities and low rate of filtration makes sedimentation an effective process.

Charged sand particles absorb particles opposite charge by electrostatic attraction. Clean quarts sand has negative charge and binds positively charged particles such as flocs of carbonates, aluminum, hydroxide and cations of iron and manganese.

Organic particles and other particles adsorbed into filter sand are subjected to microbial and biochemical activities. Organic matter is broken down by micro organisms and microorganisms after their death become substrates for other micro organisms. Organic matters are slowly transformed into inter organic compounds such as CO_2, nitrates, sulphate and phosphates which are finally discharged along with effluent.

Back washing is a process in which filtered water with air bubbles is passed into sand bed in reverse direction. In RSF the flow rate of water decrease after accumulation of solid flocks. These flocs have to be removed so that flow rate is normal during operation. Solid flocs rise up and are removed by the flow of water in reverse direction.

Disinfection of Drinking Water

Disinfection is the final process to which water is subjected prior to distribution. Various chemical agents have been used for disinfection of drinking waters.

A. Chlorination

They are less harmful, more effective and quick in action. Chlorination process may be of two types. In pre chlorination method chlorine is applied prior to any other treatment usually for controlling taste and odor. In post chlorination method chlorine is applied after treatment process especially after filtration.

Chlorination is the application of chlorine to water in small quantities either directly or as chlorine compounds. It is an oxidizing agent which can also directly react with bacteria and viruses. It even kills endospores of bacteria and fungal spores.

Mechanism of Action

Chlorine reacts with water to produce Hypochlorous acid (HOCL) and hypochlorites ($NHCl_2$) and dechloramine which are together known as combined available chlorine

$$\textit{Eg:}\ Cl_2 + H_2O \rightarrow HCl + HoCl$$

It is a strong oxidizing agent which reacts with cellular compounds particularly cell membrane and brings about death of micro organisms. When chlorine is dissolved in water it releases 4 types of compounds such as chloramines, hypochlorite ions, hypochlorous acid and singlet oxygen. The action of these compounds brings about complete destruction of micro organisms.

The mechanism of action of Chlorine depends on various factors:

- The water to be chlorinated should be free from turbidity.
- The chlorine demand of water should be estimated.
- The chlorine demand of water is the difference between amount of chlorine added to water and the amount of residual chlorine remaining at the end of specific period of contact (usually 60 m) at a given temperature and pH of water. The point at which chlorine demand of water is met is called break point. Free chlorine (HOCl and OCl) begin to appear in water.
- The presence of free residual Chlorine for a contact period of at least one hour is essential to kill bacteria and viruses. The minimum recommended concentration of free chlorine is 0.5 mg/1 for one hour.
- Super chlorination is a method which is followed by dechlorination and involves addition of large doses of chlorine after disinfection. This method is applicable to heavily polluted water whose quality fluctuates greatly.

B. Ozonization

Ozone is relatively unstable gas. It is a powerful oxidizing agent. It eliminates undesirable odors, taste and colors. It inactivates virus in few seconds. Ozone destroys virus, bacteria and organic compounds that are precursors for undesirable chloro-organic compounds that form when chlorine is added. Ozonization is usually employed in combination with chlorination.

C. UV Radiation

UV rays are effective germicidal agent against viruses and bacteria. The process is more expensive than chemical treatment. A further disadvantage is that color and turbidity reduces the disinfection potential of UV rays. UV rays are not used for large scale water disinfection.

The water which is disinfected reaches the consumer after it is stored in Storage tanks of various capacities. A flow chart representing the entire process is represented through Flow chart which clearly indicates the steps involved in the treatment process (Flow chart 1).

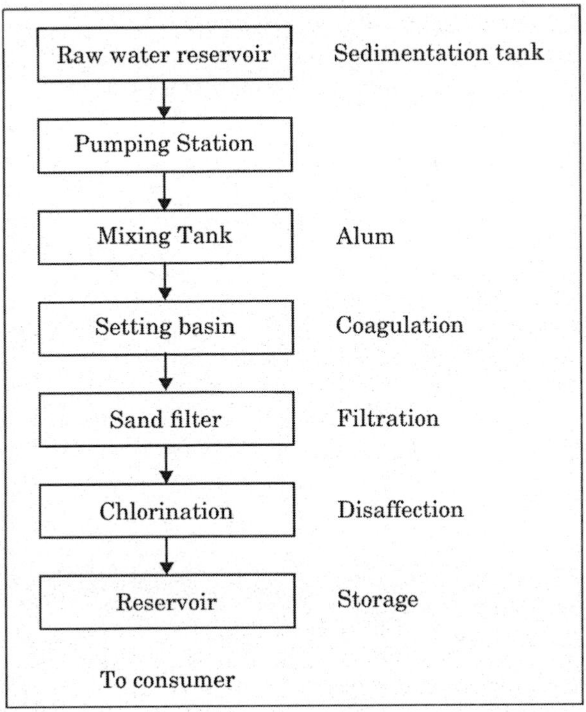

Flow chart 1: Flow chart indicating the steps involved in Water Treatment

5.5. SEWAGE MICROBIOLOGY

Human Demographic patterns of density populated areas, large scale agricultural operations and major industrial activities result in production of liquid water including waste from everyday human activities, agricultural and industrial waste which is the used water of a community otherwise called as sewage or waste water.

These liquid waste discharges flow through natural drainage patterns or sewers (to the sewage treatment plants). This result in production of liquid waste on a scale routinely overwhelms the self purification capacity of Aquatic ecosystems. Such natural bodies of water serve local communities as a source of drinking water irrigational purpose, household use fish and shellfish production, swimming or other recreational purpose. To maintain the quality of water it is essential that sewage should undergo treatment before deposal of this into natural bodies of water.

Reasonably low amounts of raw sewage can be accepted by natural waters without causing any significant decline in the level of water quality. There are mainly two groups of organisms found in water Autochthonous (indigenous) which are found indigenous to sewage degrading organic compounds in sewage. Allochthonous (non indigenous) are foreign and in club populations of pathogens and non pathogens that into aquatic ecosystem.

Sources of Waste Water

The waste water is discharged from various sources

1. ***Domestic waste water*:** Including human excrement and wash water.
2. ***Industrial water borne waste*:** Include acids, oils, greases and animal and vegetable matter discharged by factories.
3. ***Agricultural waste*:** From intensive live stock farmers where a number of cattle hogs, chicken etc are fed in small area. Pesticides from agricultural field, manure etc.
4. Surface, storm and atmosphere water.

Any of these sources as pollutants can affect the Environment and health of people (Table 9); Hence must be treated before been released into the Environment.

Domestic Waste Water

Domestic waste water include the waste water discharged from residence and commercial institutions and similar facilities include human excreta with pathogenic micro organisms (which is of particular concern) vegetable wastes and waste from washing, bathing etc. This water has high BOD and many pathogens from human sources.

Industrial Waste Water

Industrial waste water is the major source of waste water which includes water from various industries like sugar industries chemical plants, refineries, food processing industries etc. Depending on industries the composition of waste water varies. The waste water includes various carbohydrates, nitrogenous particles that are released from cannery industries, green peas etc. Chemical industries may release ammonia and beverage industries release waste products, vegetable matter paper waste etc. Industrial waste also contributes acids salts of metal and other inorganic matter. Thus the water has a high BOD and toxic compounds.

Table 9: Impacts of pollutants to Ecosysytem

	Pollutant	*Effects*	*Comments*
1.	Organic waste (animal food processing, oil refineries, paper, leather and textile plants etc	Increases BOD of water	Destroy autochthonous organisms. Decomposition becomes limited.
2.	Pathogenic Micro organisms	Diseases in humans (Water born diseases	Many bacterial, viral and protozoan diseases
3.	Inorganic chemicals and minerals	Increases salinity and acidity of water and render it toxic compounds	Mercury like chemicals are toxic to humans and lead to serious health hazards
4.	Nutrients	Causes excessive growth of aquatic plants and algae by Eutrophication and result in undesirable odors taste etc to water	Removal of phosphates and nitrates from water during waste treatment is difficult and costly
5.	Radioactive wastes	Cause birth defects and radiation sickness when in large doses	Effect can be magnified through food chain
6.	Heated water	Reduces O_2 solubility in water. Encourages growth of undesirable organisms and decreases growth of derived organisms	

Waste Water from Agricultural Source

Include the waste released from intensive live stock farming and hence contain sewage rich in organic compounds. Also contain organic compounds from agricultural field pesticides, fertilizes and other similar compound. They may contain good amount of chloride also.

Surface Storm and Atmospheric Water

Include the water collected from rainfall, snow etc. Runoff water contains various organic and inorganic compounds depending on the locality.

The waste water collected from all these sources are taken to municipal sewage treatment plants through sewage system which consist of

1. ***Sanitary sewers*:** Which carry domestic and industrial waste water.
2. ***Storm sewers*:** Designed to carry surface and storm (atmosphere) water.
3. ***Combined sewers*:** which carry all waste water though single source

Characteristicss of Waste Water

Domestic waste water or sewage consist of approximately 99.9% water and 0.02–0.03% suspended solids and other soluble organic and inorganic constituents. This consists of 70% and 30% inorganic solids.

The solid content occurs in both suspended and dissolved forms. The inorganic compounds include Ammonia, Chlorides, salts and metals. These inorganic compounds are actually which are present in water supply. Metal industries and mines also contribute to inorganic compounds. Organic compounds may be non nitrogenous such carbohydrates or nitrogenous such as proteins. Organic substance are mainly the contribution of plant and animal waste (animal sewage is high content of cellulose and lignin).

Sewage

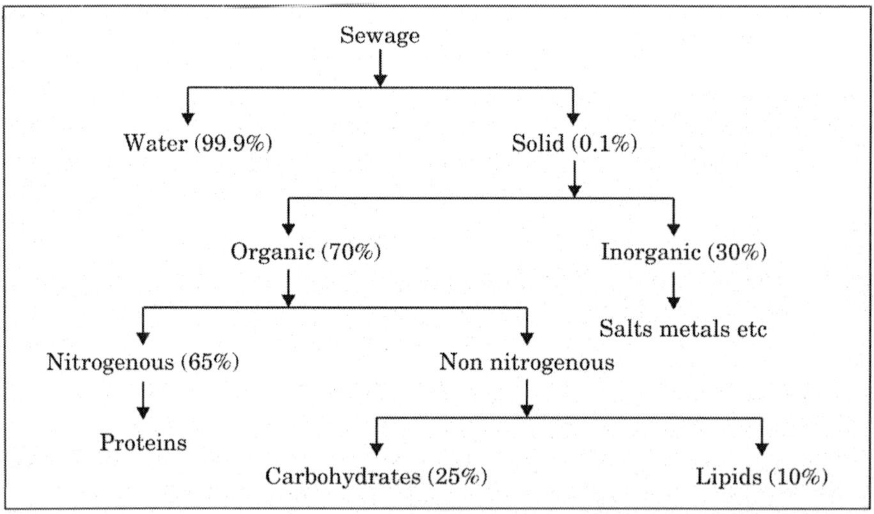

Lipids as fat and fatly acids escape digestion, absorption a deposition in human digestive systems and accounts for lipids on feces. Proteins are mainly originated from excreta of humans and animals. In addition dead/decomposing plants, insects and micro organisms also provide proteins.

Chemical Characteristics of Waste Water

Waste water is mostly water by weight. Other materials makeup only small portion of waste water but can be present in large enough quantities to endanger public health and the environment.

Organic matter

Organic matter in waste originate from plants, animals, synthetic organic compounds and enter waste water in human wastes, paper products detergents, cosmetics foods and from agricultural, commercial and industrial sources.

The principle group of organic substance found in waste water is

1. Proteins (40–60%)
2. Carbohydrates (25–50%)
3. Fats and oils (10%)
4. Urea etc

These are the waste originating mainly from plants, animals, human wastes, paper products, detergents etc. All animal and plant food stuffs contain proteins. When proteins are present in large quantities extremely foul odors are produced by their decomposition. Waste from slaughter houses meat processing industries etc will release large amount of proteins. Carbohydrate can be discharged from fermentation industries, sugar factories vegetable processing industries and include sugar, starch, cellulose, wood fibers etc. Cellulose among these is most abundant carbohydrate found in waste water.

Oil and grease

Fatty organic matter from animal's, vegetables and petroleum are not quickly broken down by bacteria and cause pollution in receiving environments. They are third major compounds of food wastes. They contribute to domestic waste water as in butter, lard, vegetable oils etc. It is also found in meat, fruits and hence related to industries involved in processing of these food stuffs.

Petroleum based waste oils used for motors and industry is considered hazardous and should be collected and disposed of separately from waste water.

Inorganic matter

Inorganic minerals, metals and compounds such as sodium, potassium, magnesium calcium, copper lead nickel and zinc are common in waste water from both residential and nonresidential sources. Most inorganic substances are relatively stable and cannot be broken down easily by microorganisms.

Large amounts of much inorganic substance can contaminate soil and water. Some are toxic to animals and humans and may accumulate in the environment. For this reason extra treatment steps are often required to remove inorganic material from industrial waste water resources.

Nutrients

Waste water often contains large amount of nutrients, nitrogen and phosphorus in the form of nitrate and phosphate which promote plant growth. Organisms only require small amount of nutrients in biological treatments, so normally there is an excess available in treated waste water. These cause algae and other plants to grow excessively depleting oxygen in thc water. Deprived of oxygen, fish and other aquatic life die emitting foul odors.

Nutrients firm waste water may cause illness in humans. Nitrogen in drinking water may contribute to miscarriages and is the case of serious illness infants called methemoglobinemia or "blue baby syndrome".

Gases

Gases commonly found in waste water include N_2, O_2, CO_2, H_2S, NH_3, CH_4 etc. which are formed by decomposition of waste matters in sewage. The gases NH_3 and H_2S can be toxic and leads to asphyxiation problems. It is also dangerous to fishes. As these gases emit odors it can lead sewers nuisance. Methane a byproduct of anaerobic digestion is also highly poisonous.

Physical Characteristics of waste water

Solids

Solid materials in waste water consist of organic and inorganic materials and micro organisms. The solids must be significantly reduced by treatment as they can increase BOD when discharged into receiving waters and provides places for micro organisms to escape disinfection

Solids are 3 types

1. ***Settleable solids*:** Certain substance such as sand grit and heavier organic and inorganic materials settle out from rest of waste water stream during preliminary stages of treatment. On the bottom of settling tanks and ponds organic material makes up a biologically active portion of sludge that aid in treatment.
2. ***Suspended solids*:** Materials that resist setting may remain suspended in waste water. Suspended solids in waste water must be treated as they will clog soil absorption systems or reduce the effectiveness of disinfection systems.
3. ***Dissolved solids*:** Small particles of certain waste water materials can dissolve like salt in water. Some dissolved materials are consumed by micro organisms in waste water but other such as heavy metals is difficult to remove by conventional treatment. Excessive amounts of dissolved solids in waste water can have adverse effects on environment.

Temperature

The temperature of waste water varies form 10–21°C. In general biological treatment activity accelerates in warm temperature and solve down in cool temperature. Extreme hot on cold can stop treatment process altogether.

Hot water as a byproduct of many manufacturing process can be a pollutant when discharged in large quantities. It can raise the temperature of receiving streams locally and disrupt he natural balance of aquatic life. Also oxygen is less soluble in warm water then in cold water. Moreover an abnormally high temperature favors the growth of undesirable organisms and stops the growth of desired organisms involved in decomposition of sewage.

pH

The acidity or alkalinity of waste water affects both treatment and environment. Low pH indicates increasing acidity while a high pH indicates increasing alkalinity. The pH of waste water needs to remain between 6 and 9 to protect organisms (Acids and other substance that alter pH can initiate treatment process when they enter waste water from industrial or commercial sources.

Odor

The gases produced during decomposition of organic matter such as H_2S (Produced during anaerobic digestion of organic waste), Ammonia etc. create odors.

Color

Fresh sewage is usually light brownish or grey in color. When anaerobic condition develops a change to dark grey and ultimately to black (due to some metallic Sulphides) happens.

Turbidity

The measurement of turbidity is based on the intensity of light scattered by sample (sewage) or compared to light scattered by control under same conditions. Colloidal matter will scatter or absorb light and prevents its transmission. Increasing turbidity of waste water increases the load of contamination.

Microbiological Characteristics of Waste Water

Since the composition of waste water varies it is to be expected that types and numbers of micro organisms will fluctuate. Fungi protozoa, algae, bacteria and viruses are predominant members.

A. Bacteria

Fecal matter contain upto 10^{12} bacteria/gram. Bacterial content of feces represents approximately 9% dry night.

Coliform group of bacteria

Include aerobe and facultative anaerobic gram negative non sporulating bacteria that produce acid gas from lactose. The classical species of the group are *E. coli* and *E aerogens*. Other members of this group belong to the genus *Salmonella, Shigella, Klebsiella, Proteus, Serratia*, Gram negative aerobe bacteria such as *Pseudomonas, Alcaligens, Flavobacterium* etc are microorganisms belonging to these groups. The pathogenic forms cause diseases and are responsible for disease outbreaks (Table 10).

Table 10: The pathogenic bacterial forms found in sewage

	Bacterial agent	***Disease***	***Major reservoirs***	***Site affected***
1	*Salmonella typhi*	Typhoid fever	Human feces	Gastro intestinal tract
2	*Salmonella paratyphi*	Paratyphoid fever	Human feces	Gastro intestinal tract
3	*Shigella*	Bacillary dysentery	Human feces	Gastro intestinal tract
4	*Vibrio cholerae*	Cholera	Human feces	Gastro intestinal tract
5	*Pathogenic E. coli*	Gastro enteritis	Human feces	Gastro intestinal tract
6	*Yesinia enterocolitica*	Gastro enteritis	Human/ Animal feces	Gastro intestinal tract
7	*Klebsiella pneumonia*	Pneumonia	Human feces	Respiratory tract

Anaerobia bacteria like *Clostridium perfrigens* are spore formers found in water contaminated with sewage containing feces of humans

Predominant physiological types of bacteria may shift during the course of sewage digestion *Eg:* in anaerobic digesters facultative types *Enterobacter, Alcaligens, Pseudomonas, Escherichia* etc, predominates during initial stages. This is followed by methane produces which are strict anaerobes *Eg: Methanobacterium, Methanococcus Methanosarcina* etc.

A predominant feature of river water receiving sewage effluent is presence of filamentous aerobic bacterium *Sphaerotilus natans* (sheathed bacteria which are efficient degraders of sewage in river water (2g/hour/square metre).

B. Viruses in Sewage

Waste water can become contaminated by approximately 140 types of entirc viruses. Many of them are pathogenic and enter directly to drinking water (swimming etc) and reach indirectly through food (shellfish, vegetables etc) and results in various diseases.

Eg: Hepatitis virus responsible for Hepatitis

Adenoviruses responsible for gastroenteritis etc. Bacteriophages infecting bacteria are also found in good numbers.

C. Protozoans

Giardia lamblia responsible for diarrhea and abdominal cramps and *Cryptosporidium* responsible for cryptosporidiosis (a disease characterized by diarrhea and abdominal cramps) are some of the important protozoans. The importance of identifying specific microbial pathogens in potable water was accelerated in Milwauke in USA with outbreaks of Cryptosporidiosis disease in 1993.

5.6. WASTE WATER TREATMENT

Single Dwelling Units

Treatment of waste as in individual homes rural areas and quite few residents of towns and small cities depend on septic tank, imhoff tank, cesspools or oxidation ponds etc. for treatment of wastes.

Role of Biological Agents in Waste Water Treatment

Biological, bacterial or other microbial cultures break down organic compounds and remove organic matter. Bacteria's, Protozoans etc play very important role in biological treatment of sewage in single dwelling units.

Septic Tank

Septic tank method is also known as anaerobic digestion systems. In this, small or large quantities of sewage water from individual units or houses can be carried out. This unit accomplishes two processes.

1. Sedimentation
2. Biological degradation of sewage

A septic tank is an enclosed concrete box into which waste water flows from the source. The sewage after it reaches the septic tank is retained for a short time. Sedimentation occurs from the upper position permitting liquid with fewer suspended solids to get discharged from the tank. Thus the organic material in the sewage settles to the bottom of the tank where it is covered by a thin organic film which excludes oxygen.

The organic matter are hydrolyzed and fermented by anaerobic bacteria. The end products are very unstable and has high BOD and odor. The anaerobic bacteria present digest sludge by converting organic matter into simple chemical compounds or gases.

The gases produced during anaerobic decomposition are discharged through a vent in the tank as the clear water (after sedimentation) passes through series of buried perforated tubes. The effluent percolates into the soil where the dissolved organic compound undergo biodegradation.

It is a small scale anaerobic treatment process and is simple inexpensive and satisfactory design if properly operated. But various problems associate with usage of septic tank which include

1. Septic tank treatment does not reliably destroy intestinal pathogens and hence leakage of pipes or drainage of the effluent results in contamination of drinking water supplies (If drinking water supply is in close proximately to septic tank)
2. The end products are still unstable and high in BOD and odors.
3. Undigested solids in the bottom of septic tank must be periodically removed to prevent clogging of pipes.

Water enters the enclosed septic tank where solids fall to the bottom and accumulate as sludge which in turn decomposes biologically. Relatively clear water basses through outlet pipe and is distributed by distributor box into buried field of clay pipes that open into soil.

Imhoff Tank

Imhoff tank was designed by Karl Imhoff of Germany. An imhoff tank is an improved septic tank in which the incoming sewage of influent is not allowed to get mixed up with sludge produced and outgoing effluent is not allowed to carry with it large amount of suspended water (as in case of septic tank).

It is on larger scale compared to septic tank and is used for treatment of sewage of community or small city etc. It is built in two storage with two compartments- upper and lower compartments.

1. ***Upper chambers*** – is called sedimentation or flow through chamber through which sewage flows at a very low velocity
2. ***Lower chamber*** – in which anaerobic or septic digestion (or decomposition) occur. The solid of the sewage settle to the bottom of sedimentation chamber where digestion take place. They are made to fall to digestion chamber through an entrance slot at lowest part of the sedimentation tank (chamber)

The sedimentation tank is in way that gases generated in the digestion chamber cannot stay in sedimentation chamber. The gas vent collects the gases escaping from the surface. The chief gas is Methane having a considerable fuel value and collected separately for use if required.

The digestion zone (chamber) is made up of 2 or 3 inverted zones called hoppers with sides opening one to one so as to concentrate sludge at the bottom of the hopper. The sludge is periodically removed through sludge pipe, the flow lining under a hydrostat pressure. All sludge is not removed. Only the lower layer decomposed are withdrawn to keep the tank secured with anaerobic bacteria to permit uniform distribution of settled solids throughout the length of digestion chamber so as to utilize storage capacity in the greatest measure.

Sewage slowly enters chamber, solid matter settles out and falls through bottom and collects in lower chamber where anaerobic digestion occurs.

Cesspools

In many homes human wastes is dumped into cesspools consisting of concrete rings (cylindrical rings) with pores present in it. After the solid waste accumulates in the bottom, anaerobic decomposition take place. The liquid passes through pores into surroundings sand where further decomposition take place.

Waste enters through inlet pipe and solid material falls to the bottom of cesspool to form sludge. Water passes through open bottom and through pores in the wall of cesspool into surroundings soil.

Micro organisms involved in there processes are

***Bacteria*:** *Pseudomonas, Nitrosomonas, Zooglea, Sphaerotilus, Baggiatoa, Achromobacter, Flavobacteria, Nocardia, Escherichia, Bacillus* etc.

***Fungi*:** *Zoophagus, Geotrichum, Alternarda, Pullularia, Penicillium, Candida, Cephalosporium*

***Protozoa*:** *Phytoflagellates, Zooflagellaltes, Amoebae* and *Ciliates.*

5.7. MUNICIPAL WASTE WATER TREATMENT

The primary objective of sewage treatment is the removal and degradation of organic matter under controlled conditions. Municipal plants are equipped for a mechanic sewage treatment and handles massive amount of daily generated waste and garbage.

Based on the nature of process involved differed stages can be grouped into

1. ***Physical processes***: These processes depend on physical properties of impurities like particle size, specific gravity, viscocity etc. *Eg:* Screening, sedimentation and filtration
2. ***Chemical processes*:** These processes depend on chemical properties of either the impurities or the added reagents.

   Eg: coagulation, precipitation and ion exchange.
3. ***Biological process*:** There processes utilize biochemical reactions to remove soluble colloidal organism. The process may be aerobic or anaerobic.
4. ***Aerobic processes*:** include biological filtration, oxidation ponds and activated sludge process.
5. ***Anaerobic process*:** includes sludge digestion.

Sewage Treatment Involves Three Stages

1. ***Primary treatment*:** Involves removal of suspended solids or settleable solids with 30–40% reduction in BOD.

   Eg: Screening, coagulation and sedimentation
2. ***Secondary treatment*:** Organic matter in sewage is oxidized by micro organisms and 90% reduction in BOD is achieved.

   Eg: Tricking filter, activated sludge process and oxidation ponds.

3. ***Tertiary treatment:*** Where inorganic salts, organic and inorganic suspended and dissolved solids and poorly biodegradable organic material are removed including toxins.

   Eg: Anaerobic sludge digestion, ion exchange process etc.

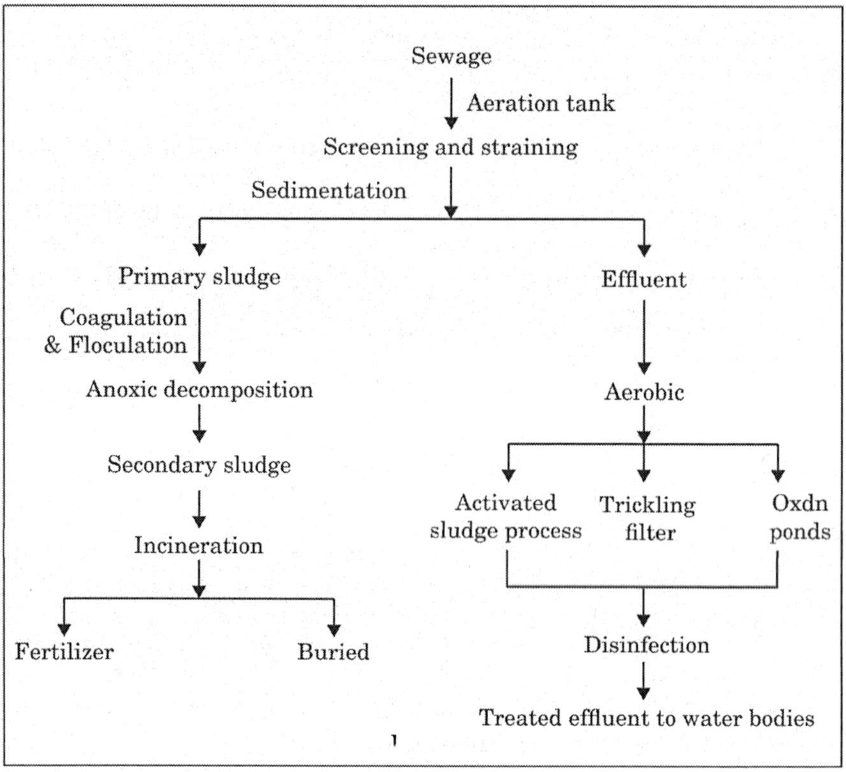

Flow Chart 2: Sewage Treatment

Primary Treatment /Preliminary Treatment

To protect the main units of treatment plant and to ensure their efficient operation etc is necessary to remove large floating and suspended solids like leaves, twigs, paper and other debris from waste water. This can be achieved by preliminary treatment processes like

1. Screening
2. Coagulation
3. Sedimentation

Screening

The first stage of preliminary treatment process involves screening to remove large. Coarse screens with openings of about 75 mm are used to present large objects. Main screens are provided with openings of 5–20 mm through which incoming water passes.

To present overloading of main screen with collected solids, the screen is slowly rotated so that the solids can be removed at regular intervals.

Screenings from sewage are usually disposed of by burial or incineration. Alternatively they may be passed through maceration, which shreds them to small size. Then they are retained to the flow so that they can be removed along with settleable solids during the main treatment process.

Micro staining

Micro-strainers is a drum screen with fine woven stainless steel mesh. This can be used in final testing stage to produce high quality sewage efficient.

Grit Removal

Considerable amount of grit is carried along with flow of sewage. If not removed these grit particles may damage the mechanical parts of treatment system. Grit particles are of large size and have high density when compared with organic particles in sewage. Thus they can be removed using principle of differential settling.

Grit particles with a diameter of 0.20 mm and specific gravity of 2.65 settle at a velocity of about 1–2 n/min whereas responded solids have considerably lower settling velocity with a retention time of 30–60 seconds, the flow of sewage through channel of sufficient length can alone result in settling of grit particles at the bottom while remaining suspended solids continue to flow through the channel. The settled grit is removed at intervals, washed and then disposed off.

Coagulation

Many impurities in waste water are present as colloidal solids. Sedimentation can be used to remove suspended particles down to a size of about 50 mm depending upon density. Smaller particles have very slow setting velocities and cannot be removed by sedimentation. These smaller particles can be removed once they are allowed to agglomerate. Agglomeration by flocculation with or without the use of a coagulant may lead to an increase in size of particle and particles are then removed by sedimentation or flocculation.

Coagulation is the process of agglomeration of smaller particles by adding a coagulant. The coagulants are a metal salt which reacts with alkalinity in water produce an insoluble metal hydroxide floc which precipitate the colloidal particles. This fine precipitate is then flocculated to produce settleable solids. That is larger flocs created through agglomeration enhances the separation and helps settle down at a faster rate.

The coagulant is usually required in doses of 30–500 mg/l. the commonly used coagulant is Aluminium sulphate commercially called as Alum. The overall reaction includes.

$$Al_2(SO_4)_3 + 3Ca(HCO_3)_2 \rightarrow 2Al(OH)_3 + 3CaSO_4 + 6CO_2$$

The solubility of $Al(OH)_3$ is pH dependant and is low between 5 and 7.5. Other coagulants used include Ferrous sulphate ($FeSO_4 . FH_2O$) Ferric sulphate ($Fe(SO_4)_3$) and Ferric Chloride ($FeCl_3$)

Coagulant has to be despaired throughout the body of water before flocculation occurs. A rapid mixing chamber with a high speed turbine can be used for this purpose.

Sedimentation

Sedimentation or classification is the setting and removal of suspended particles that take place when water stands still (or flows slowly through a tank). When water is still particles that have density higher than that of water will settle down under the influence of gravity. The settling particles form sludge layer at the bottom of the tank and classified water will be collected through the outlet.

Types of Sedimentation Tanks

The main types are

1. Rectangular horizontal flour tank
2. Circular radial flour tanks.
3. Vertical flour tanks.

Rectangular horizontal flour tanks are most commonly used ones. They are normally built of bricks or concrete and is rectangular in plan and cross section. The inlet for raw circular and outlet for clarified water are made at opposite end of tanks. In between inlet and outlet is the setting or sedimentation zone. The sedimented materials form a sludge layer at the bottom of the tank. To facilitate sludge removal, tank bottom is built in such a way that it stops slightly towards the inlet where the opening for sludge collection is situated. The sludge has to be cleaned regularly. This can be done by manual cleaning in which the tank must be drained first. Alternately the bottom of the tank can be provided with a continuous belt system with screen.

Efficiency of Sedimentation Tank

The sedimentation velocity (Vs) of a particle can be determined by calculating the depth of the tank (H) it traverses during the detention time (T).

$$Vs = \frac{H}{T} \text{ (m/hr)}$$

Detention time can be calculated as

$$T = \frac{Alt}{Q}$$

A – surface area of tank (m^2)

H – depth of tank (m)

Q – influent flow rate (m^3/hr)

When sedimentation is used without pretreatment as in case of clarification of river water, the process is called plain sedimentation.

Secondary Treatment

The effluent from primary sedimentation tank still contains a proportion of organic matter in solution or colloidal state and numerous micro organisms. Its BOD is high, hence subjected to aerobic oxidation by various methods.

It is necessary to oxidize the organic matter present in effluent before it is being discharged into the body of water (dilution). Secondary treatment relies on microbial actively. Since those oxidation processes are carried out by micro organism they are called as biological oxidators.

There are four different types of oxidation process.

1. Fixed film sewage treatment
   a. Biological filters (Trickling filters)
   b. Rotating biological contractor
2. Suspended cell sewage treatment
   a. Oxidation ponds
   b. Activated sludge process

A. Trickling Filter (Bacterial bed/Percolating filters)

Trickling filter is the oldest form of aerobic oxidation treatment. It is a simple method but expensive, which works on the principle of filtration.

Construction

It consists of a circular bed of stone, gravel or synthetic material with a particle size of 50–10 cm. The filter bed is usually 1.5 to 2.0 m depth and is supported by a layer of larger stones over the under drains. There is a rotating arm above the surface of filter bed with nozzles. The rotating arm sprays the settled sewage over the filter bed either continuously or intermittently. The spraying saturates the liquid with oxygen.

Within the interstices of medium the micro organisms grow in the form of a slime film. The film is often known as zoogleal film and consists of bacteria, fungi, protozoa and algae. A newly constructed bed must acquire zoogleal film before it can function efficiently.

Working Principle

Effluent sprayed over the bed slowly trickles through the interstices on the medium which is coated with zoogleal film (Liquid flows over the film than through it). When the liquid flows over microbial film the organic material bind to it and are oxidized. The microbes here are stationary portions of films and may be sloughed off by shear forces created by flow of waste water. The treated effluent containing the sloughed off biomass leaves the bed *via* the under drain into sedimentation tank. This tank is known as humus tank and sludge which sediment at the bottom of the tank is called as humus sludge.

The highest rate of oxidation take place in upper regions of bed, where oxygen is supplied by natural ventilation. Below this level, rate of oxidation decreases due to decreasing concentration of organic matter in liquid phase.

Microorganisms involved

The upper region of trickling fitter where light is available to support photosynthesis is dominated by algae of genera *plasmodium*, *chlorella* and *ulothrix* (At times growth may be so extensive that it impairs the operation of filter).

Zoogleal film is developed due to presence of filamentous bacteria like *Sphaerotilus natans* and *Baggiatoa*. Other bacterial genera which oxidize organic matter are *Pseudomonas*, *Flavobacterium*, *Achromobacter*, *Alcaligens* etc. In the lower layer nitrifying organisms like *Nitrobacter* and *Nitrosomonas* predominates.

Many species of fungi and protozoa occur throughout the filters, the number of which is influenced by availability of oxygen and nutrients. Fungal genera include *Fusarium, Penicillium, Mucor, Geotrichum, Sporotrichum*. Among protozoa the predominant group is Ciliates and include *Opercularea, Verticella* etc. Other protozoa forms are *Amoeba, Euglena, Trichomonas, Paramoecium* etc.

Drawbacks

Nutrient overload may lead to excess microbial slime reducing aeration and percolation rates and lead to removal of trickling filter led. Trickling filter method cannot be used during cold countries when temp is very low since the growth rate of micro organisms become very low.

B. Activated Sludge Process

This is the most common option in secondary treatment. It starts with aeration that encourages the growth of microbes in waste. Those microbes in the firm of flocs which is kept in suspension by agitation feed on organic material. Agitation is provided either by mechanical means or by aeration. In this process a portion of (20%) separated sludge along with native population of living micro organism is added to incoming effluent as inoculums. This added sludge is often referred to as Activated sludge and carries out actual oxidation. This constant microbial population is maintained in activated sludge tank.

Construction

The activated sludge tank is simple in design. It is an oblong deep tank provided with an inlet at the top of one end and outlet at the bottom of other end. Aeration is provided either by an air defuser located of bottom of tank or by agitator at the surface of wastes along both sides of the tank.

Working Principle

During the process to effluent water from primary sedimentation tank is mixed with a portion of activated sludge from final settling tank. The mixture of biomass and waste water is termed mixed liquor (and the Concentration of biomass on tank as mixed liquor suspended solids mass). The mixed liquor is thoroughly aerated by agitators. This mixed liquor is maintained in suspension by turbulance created by agitators (In absence of agitation the solids quickly settle to the bottom and lose contact with organic matter in liquid stage).

- Sufficient air is transferred to mixed liquor to maintain dissolved oxygen 1–2 mg/litre
- After a period of 4–8 hrs the contents are transferred to final settling tank
- In final settling tank, the microbial biomass along with other suspended solid sediments as sludge. This sludge is usually referred to as secondary sludge (to differentiate it from sludge of primary treatment)
- The supernatant is discharged as effluent

Principle

In activated sludge process biomass exist as freely suspended flocs in which the cells are embedded in gelatinous metrix. This basic unit of activity of the process is floc, which intern is dependant for its existence on the metabolism, growth and physical properties of microbial cells. The successful operation depends on

- Ability of floc to absorb substrates
- Assimilation and oxidation of organic matter
- Oxidation of nitrogen
- Maintenance of good flocculation

Micro-organisms involved

Micro organics are maintained at high levels by introduction of most of settled activated sludge (also called returned sludge) which is rich in culture of aerobic bacteria. Even though various micro organisms are involved, the most important one is rod shaped *Zooglea ramigera* which produce a copious extracellular slime matrix. This organism is the main agent of flocculation. Among other bacterial forms *Pseudomonas, Nitrosomonas, Flavobacterium, Alcaligens, Brevibacterium, Bacillus, Baggiotoa, Achromobacter, Corynebacterium and Sphaerotilus* predominates. Other minor genera include *Aerosomonas, Aerobacter, Micrococcus, Spirillum, Acinetobacter, Hyphomicrobium* etc

Among fungi *Zoophagus, Geotrichum, Pullularia, Alternaria, Penicillum, Cephalosporium* etc. have been noted. In protozoa dominant ciliates are *Verticella, Opercularia, Aspidisca* etc. In addition yeasts and algae are of rare occurrence.

Advantages

Requires less space the final efficient of high quality is produced such that it does not require high dilution for disposal.

Disadvantages

Since the incoming effluent is introduced, at one end of the tank BOD value will be higher. Moreover micro organisms at this end will be physiologically more active than those at other end.

C. OXIDATION PONDS/LAGOONS

They are used for simple secondary treatment of sewage effluent. Within an oxidation pond heterotrophic bacteria degrade organic matter in sewage, which results in production of cellular material and minerals which supports the growth of algae. In oxidation pond growth of algal populations allows the decomposition of organic matter by the production of oxygen. The production of Oxygen replenishes the oxygen used by heterotrophic bacteria. The use of oxidation ponds are largely restricted to warmer climatic regions because they are strongly influenced by seasonal temp changes.

Types of Oxidation Ponds

Based on their activities oxidation pounds can be grouped in to four

- Facultative ponds
- Maturation pond
- Anaerobic pond
- Aerated ponds

1. ***Facultative ponds*:** Oxidation pond in which both aerobic and anaerobic activities are carried out in a single unit is called *facultative pond*. These are most widely used oxidation ponds mostly rectangular in shape with a length breadth ratio of 3:1 and 1–2 m deep. The normal retention time for these ponds is 5–30 days depending upon temperature.

   Near the surface of ponds chlorophyll bearing micro organisms phytoflagellates and algae dominate. Genera of algae include *Chlorella, Chlamydomonas, Scenedesmus, Euglena, Oscillatoria* etc. These organism perform photosynthesis and causes an increase in dissolved O_2 (DO) level, which favors aerobic bacteriological activity. Members of aerobes include *Pseudomonas, Achromobacter, Flavobacterium* etc. In the bottom region anaerobic condition maintained decompose organic materials. The anaerobic activity stabilizes sludge to some extend and release some organic matter in soluble from which are degraded in aerobic zone.

   Removal of bacteria occurs by natural death and also by settlement, because of long retention time in tank. Bacteria and Phyloplankton can also be removed by preying action of ciliates, rotifers, etc. However bacteria may escape these and can be found in the effluent.

2. ***Maturation ponds*:** This types of ponds are used during secondary stage of treatment following a facultative pond or other biological treatment units. These are shallow ponds and can be loaded with only a limited amount of organic material.

   Heavy algal growth is a problem with this type of pond nevertheless they are efficient in removing pathogenic bacteria

3. ***Anaerobic ponds*:** There are comparatively deep ponds with a depth of 3–5 m to ensure anaerobic conditions. They can be operated with a high organic loading. Retention time for these pond is about 30 days. They are generally used for pretreatment

of strong organic wastes before adding them to facultative ponds. The main problem with anaerobic ponds is productive of odor.

4. ***Aerated ponds***: Aerated ponds are less common use when compared to other oxidation ponds. They utilize floating aerators to maintain dissolved oxygen levels and to provide mixing. Retention time for these ponds is a few days. The process involves bacterial floc and produces good quality effluents.

Tertiary Treatment

All preliminary and secondary treatment reduces BOD levels of sewage. The aim of tertiary treatment is to remove non-biodegradable toxic organic pollutants such as chlorophenols, polychlorinated biphenyls and other synthetic pollutants. Various methods are employed to remove these compounds.

- Chlorophenols, Polychlorinated biphenyls etc are removed by activated biphenyls
- Phesphate is removed by precipitation as Calcium phosphate
- Nitrogen is removed by volatilization as Ammonia
- Removal of heavy metals like mercury, lead chromium, cadmium etc can be concentrated by bacteria (bacteria concentrate heavy metals in biomass)

Any escaped bacteria's or other pathogens can be destroyed by chlorination.

$$Cl_2 + H_2O \rightarrow HOCl + HCl \qquad \text{(Hypochlorous acid)}$$

$$CaO(Cl_2) + 2H_2O \rightarrow 2HOCl + Ca(OH_2) \qquad \text{(Calcium hypochlorite)}$$

The Hypochlorous acid is strong oxidant

Reverse Osmosis

Reverse osmosis is a filtration method that removes many types of large molecules and ions from solutions by applying pressure to the solution. The result is that the solute is retained on the pressurized side of the membrane and pure solvent is allowed to pass to other side. This process allows removal of salts, hardness, microorganisms and other impurities in order to improve its color and other properties.

The membranes used for reverse osmosis have a dense barrier layer in a polymer matrix where most separation occurs. In most cases the membrane is designed to allow only water to pass through this dense layer, while preventing the passage of solutes (such as salts, ions etc.). This process requires that a high pressure be exerted on high concentration side of membrane (usually 2–17 bar or 30–250 psi) .This process forces solvent from a region of high solute concentration through a semi permeable to a region of low solute concentration by applying a pressure in excess of osmotic pressure.

Osmosis describes how solvent moves between two solutions separated by a semi permeable membrane to reduce concentration difference between the solutions. The semi permeable membrane in between two solutions does not allow the solutes to move from one compartment to other, but allows solvent to move. When the solvent moves away from low concentration areas, it causes these areas to become more concentrated, on the other side when solvent moves into areas of high concentration, solute concentration will decrease. This process is termed osmosis. In reverse osmosis, pressure is applied to the compartment with high concentration. In Reverse osmosis process

- Water flows from a low dissolved solid content to a high dissolved content area (column)
- Osmotic pressure is the pressure that is used to stop water from flowing through in order to create balance
- By pursuing pressure that exceeds osmotic pressure, the water flows from the column with a high dissolved solid content to the column with a low dissolved content.

OSMOSIS – Water flows from low concentration of salta to higher concentration

EQUILIBRIUM – Osmotic pressure is the pressure required to stop water flow and reach equilibrium

REVERSE OSMOSIS – By applying pressure greater than osmotic pressure, flow of water is reversed. Water flows from higher concentration to lower concentration.

Dialysis

Electro dialysis is a membrane separation process driven by voltage where salt is moved through membrane using electrical potential thereby having fresh water behind as a product. Electro dialysis depends on principles

- Dissolved salts in water are mostly ionic *i.e.*, they are either positively or negatively charged.
- These ions are attracted towards the opposite charged electrodes.
- The charged particles can be either allowed or prohibited to pass through the membrane.

Electro dialysis is used to transport salt ions from one solution through ion exchange membrane to another solution under the influence of an applied electrical potential difference. This is done in a configuration called electro dialysis. In practical electro dialysis process multiple electro dialysis cells are arranged into configuration called electro dialysis stack, with alternating union and cation exchange membranes forming multiple electro dialysis cells.

Ion exchange membranes used for electro dialysis are their films of polymeric chains containing electrically charged functional sites. These selectively charged membranes can separate ions; if the membrane is positively charged (*Eg:* Quarternary Ammonium groups) only anions will be allowed through it and is called Anion exchange membrane, similarly negatively charged membranes (*i.e.*, with sulfonate group) are called cation exchange membranes. These are membranes that allow only mono valent cations through them.

The water (with salts or similar impurities) when fed in to electro dialysis stack through delicate circuit when arrives the active area of cells, the DC voltage causes positively charged cations to migrate towards cathode and negatively charged anions to migrate towards anode. When the ion reaches an anion exchange membrane, the membrane properties determine whether the ions is rejected or allowed to pass through. The ions that are allowed to pass through membrane are retained in next compartment since the next membrane in its path will be of opposite charge. Therefore there are compartments from where the ions are removed and some where they are concentrated.

Ion Exchange

Certain natural materials have the property of exchanging one ion their structure for another in the solution. Various types of Ion exchangers are explained in Flow chart 3.

Eg: Zeolites which are complex alumino-silicates and green sands

Higher exchange rate is possible with synthetic ion exchange materials. When used for water softening, sodium ions in the natural Zeolite are exchanged for Calcium and Magnesium ions in water. The complete removal of hardness occurs.

$$Ca^{++} \text{ or } Mg^{++} + Na_2X \rightarrow CaX \text{ or } MgX + 2Na^+$$

Where Na_2X represents Zeolite

Thus the finished water is in high in Sodium and this will not create any problems. Zeolite can be regenerated using a salt solution which provides a high concentration of Sodium ions to reverse the exchange reaction. In this case hardness is released as concentrated chloride

$$CaX \text{ or } MgX + 2NaCl \rightarrow Na_2X + CaCl_2 \text{ or } MgCl_2$$

Hydrogen ion cation exchangers

This gives water of zero hardness. They exchange all cations for hydrogen ions.

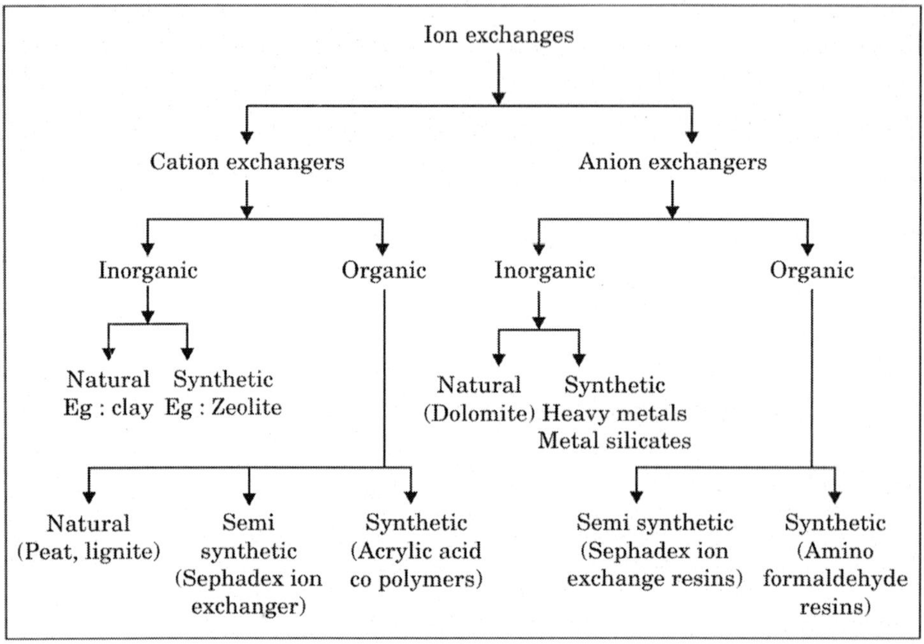

Flow chart 3: Ion Exchanges

Anion exchangers

They are usually ammonia derivatives. Anion exchangers can regenerate using a strong base.

Water of highest purity can be obtained such water is obtained using ion exchange process where anions can be replaced with hydroxyl ions using anion exchange resins or cations are replaced with hydrogen ions using cation exchange resins. The hydrogen ions and hydroxyls recombine producing water molecules. Thus no ions remain in produced water the purification process is usually performed in several steps with mixed bed icon exchange process purify metals including reporting uranium plutonium the rum etc.

Important Terminologies in Waste Water Treatment

COD (Chemical Oxygen Demand)

COD analysis is the measurement of the oxygen depletion capacity of water sample contaminated with organic waste matter. Specifically it measures the equivalent amount of oxygen required to chemically oxidize organic compounds in water.

Biochemical Oxygen Demand (BOD)

Also referred to as Biological oxygen Demand is performed to measure the potential of waste water and other waters to deplete the oxygen level of water. It is the amount of dissolved oxygen needed by aerobic micro organisms in a body of water to break down organic material present in given water sample at specific time period.

5.8. ANAEROBIC DIGESTION

Large scale anaerobic digesters are used for appropriate processing of sewage sludge produced by primary and secondary treatments. They are used for processing of settled sewage sludge and for treatment of some very high BOD industrial efficient

Anaerobic digestion is a process of stabilization of solid wastes and production of methane and CO_2 in an air tight cylindrical tank or digester. Anaerobic digester is a tank where the process is carried out. It can also be carried out at a slower rate on an unmixed plastic sheet. The heat value of methane is the same as natural petroleum gas and

Biogas and is valuable as an energy source. The digester is maintained at 35°C for combusting to induce Biogas production.

The solids which accumulate from the sedimentation tank and primary and secondary sludge are treated further in anaerobic digestion tank since biological reactions are carried out by microorganisms. The microbial action leads of sludge digestion and biogas production.

Anaerobic sludge digestion tank is specially designed tank which is maintained free of oxygen. It may be a completely closed tank or may be provided with lid at the lop, the contents are introduced in to tank though an inlet near the top. There is an out let near to bottom to remove the digested sludge. There is also a vent in the lid through which any gas produced closing the digestion escapes.

Operational Parameters

The degree of maxing required is much less than in aerobic process. Sludge digestion is a slow process and frequently requires heating to accelerate the bio chemical reactions involved. The majority of conventional digesters are operated in mesophilic range between 25 and 40°C. Neutral pH is required for sludge digestion and the retention time for mesophilic and thermophilic digesters is in the range of 25–35 days (and in some cases 12–15 days).

Anaerobic biochemical reaction takes place in the anaerobic tank. Both anaerobic and facultative bacteria degrade organic solids by converting them into soluble substances and gaseous products. The gases consist mainly of methane (60–70%) and CO_2 (20–30%). In addition H_2 and H_2 are also produced during sludge digestion in small amounts. These gas mixture can be used for heating and also for operating power. The supernatant liquid portion, the effluent can be sent to the aerobic treatment process. The digested sludge is disposed off proper way.

Micro-organs Involved

The digestion process is carried out by bacteria. Among the non methanogens *Bacillus, Pseudomonas Lactobacillus Spirillum, Vibrio Klebsiella, Alcaligens* are facultative anaerobes. *Clostridium, Bacterioides* etc. are obligate anaerobes. Methanogens include *Methanococcus Methanosarcina, Methanospirillum* and *Methanomicrobium*.

The formation of flammable gas methane is the most important process in anaerobic digestion. These are four different stages in the formation of methane through anaerobic digestion

- Hydrolysis
- Fermentation
- Acetogenesis
- Methanogenesis

1. ***Hydrolysis:*** Since large insoluble polymers cannot be directly transported in to microbial cells there are extracellular reactions mediated by bacteria which produce enzymes necessary for degradation of large molecules

   Cellulose hemicelluloses and starch comprise the important polysaccharide in anaerobic digestion process. In addition pectin and dextrin are also found to be present in digestion process. These complex polysaccharide are hydrolyzed by cellulolytic bacteria, amylolytic bacteria and pectinolytic bacteria which are able to produce the corresponding enzymes. Preteins present are degraded by proteases to peptides and amino acids
2. ***Fermentation*:** Among various substrates amino acids and sugars can be fermented readily. Acidogenic fermentation are the most important in anaerobic digestion. In the fermentation of amino acids, Pyruvate is the most important intermediate from which lactate, propionate, butyrate, formate and acetate are formed.
3. ***Acidogenesis*:** Aectate formation is an important step in anaerobic digestion because acetate can be directly converted in to methane.
4. ***Methanogenesis*:** Although acetate is an important substrate, only a few bacteria can generate methane from it (Acetoclastic bacteria)

$$CH_3COOH \rightarrow CH_4 + CO_2$$

Eg: Methanosarcina barkerii. This organism can also use method (CH_3OH) as substrate

$$CH_3OH + H_2 \rightarrow CH_4 + H_2O$$

Energy for these reactions are obtained from oxidation of molecular Hydrogen using CO_2 as terminal electron acceptor. Thus H_2 produced during formation of Acetate from Ethanol and CO_2 produced during generation of methane from acetic acid are combined to produce methane

Mechanism of Methane formation

$$CO_2 + H_2O \rightarrow CO_2 + H_2O$$

$$CO_2 + 4H_2 \rightarrow CH_2 + 2H_2O$$

$$4CH_3OH \rightarrow 3CH_4 + CO_2 + 2H_2O$$

$$4HCOOH \rightarrow CH_4 + 3CO_2 + 2H_2O$$

$$CH_3COOH \rightarrow 12CH_4 + 12CO_2$$

Factors affecting methane formation

- ***Slurry*:** The ratio between solid and liquid should be 1:1
- ***Seeding*:** Seeding of slurry activates methane evolution. It contain acetogenic and methanogenic bacteria
- pH between 6–8
- ***Temp*:** Fluctuation in temp reduces methane formalin
- ***Mesosphiles*:** - 30–40°C
- ***Thermosphiles*:** 50–60°C
- ***Carbon–Nitrogen ratio*:** Inappropriate Carbon–Nitrogen ratio lowers methane production. It should be 20:1 ratio.
- ***Addition of algae*:** The amount of Biogas produced from algae will be twice when compared to crowdung. It holds promise to get Biogas in efficient amounts.
- ***Creation of anaerobic environment*:** Anaerobic environment can be created by air tight digestion tank buried in soil.

Biogas

Biogas production is an anaerobic digestion process carried out in an air tight cylindrical tank called digester to produce methane from feedstock material (cattle dung, sludge etc.).

It is a combination of 60% Methane and CO_2 produced aerobically by biodegradation of organic matter using methanogenic bacteria.

A digester is made up of concrete, bricks, cement or steel It has a side opening called inlet to which the sludge material for digestion are incorporated into cylindrical container and a vent is present above to digesters to collect the gas after 50 days. Sufficient gas is produced in a tank which can the used for household purpose. Usually digester is burred in soil in order to benefit from insulation provided by soil.

The process takes place in 3 stages

1. ***Solubilization:*** It is initial stage where feed stock is solubilized by water and enzyme. The feedstock is dissolved in water to make slurry. The complex polymers are hydrolysed to organic acids, alcohol etc. by hydrolytic methanogens.
2. ***Acidogenesis:*** Facultative anaerobes and H_2 producing acidogenic bacteria convert simple organic material *via* oxidation/redox reactions to produce reactions such as H_2 and CO_2. These substances will serve as food for final stage.

   Fatty acid is converted in to acetate H_2 and CO_2

   Complex organic matter → Fatty acids/Alcohol → Acetate+ H_2
3. ***Methanogenesis:*** This is last stage of anaerobic digestion scheme where acetate, H_2, and CO_2 are converted by Methanogens to Methane CO_2, H_2O and other products.

Benefits of Biogas

- Biogas is mainly used for cooking and lighting purpose. Sludge is used as fertilizer
- It is used in internal combustion engines to power pumps and electric generator
- Production of energy (heat, light, electricity)
- Transformation of organic waste into high 'quality fertilizer
- Minimizes of environmental pollution
- Production of Methane gas

Composting

Composting is an anaerobic decomposition process involving production of dark brown, granular, humus like and product (compost) after thermopile biodegradations of organic matter. The primary objective of composting is to convert an unstable offensive material into stable end product.

Substrates for composting include a range of materials from municipal refuse to sewage sludge, paper garbage, agricultural wastes like leaves human fecal matter etc. For producing a stable compost organic substrate must contain adequate nitrogen. A reasonable carbon nitrogen ratio is about 50. The process involves five major stages.

a. ***Salvage*:** Domestic as well as municipal refuse contain non biodegradable materials like glass, ceramic, plastic and metals in addition to other biodegradable organic materials. Large objects are removed by mechanical screening. Ferrous materials can be removed magnetically. There are various other methods which can be practiced based on the type of waste.

b. ***Grinding, pulping and homogenization*:** For effective composting substrate should be finely divided into granular material, which permits aeration mixing and invasion by degradative microbes. Particle size of 7 cm will be suitable. A moisture level of 40–50% is ideal for composting (Low moisture content may inhibit microbial action). The required moisture content can be attained by adding water. The sludge used may be raw, dewatered, digested or a combination of these types and the nature of resulting compost will depend on sludge used.

c. ***Biological degradation*:** Commonly used method for composting involves piling up of substrates in to large heap. Bulking agents like large wood chips or barks can be mixed with substrate to facilitate aeration. After the required period of composting (about 21 days), bulking agents are separated from compost and compost is ready for use

d. ***Windrow composting*:** Heaps resembling pyramids of compost is commonly used method. Windrows are 3 cm high with a base of 9 m width. It is done in open air (so disadvantageous that it is scientifically affected by local climate).

e. ***Dano process*:** This employs drums (30 m length and 5 m diameter) which rotate slowly to facilities mixing and aeration. This involves mechanical mixing and aeration inside closed digesters. hence advantageous over windrow composting and can be completed in shorter period.

f. ***Curing and drying*:** The compost must be stabilized completely before it is added to the soil (otherwise it will remove Nitrogen which is available to plant from soil in the final stages of degradation)

   In windrow composting curing for a period of two weeks in shallow piles is done and in mechanical composting curing periods can range up to three weeks or may be absent.

g. ***Finishing*:** Compost will be mechanically dried to moisture content of less than 30% for agricultural or horticultural use. Rescreening and grinding of compost may be done to remove large piece of debris. It may be pelletized to kill pathogens before selling to the farmers.

Microbiology of composting

Although decomposition is mainly carried out by thermophilic micro organisms, the initial process is carried out by mesophilic micro organisms. The biodegradable materials such as proteins, carbohydrates etc are decomposed naturally by mesophilic microorganisms. The degradation products are usually simple organic acids which causes a drop in pH (from neutral pH). As initial reactions proceed, heat is produced which rises temp within the heap. The insulating effect of heap further causes an increase in temperature. As the temperature increase, the activity of mesosphilic microbes is reduced and finally at about 60°C mesosphiles are killed by high temperature (They may become source of nutrients for microbial forms like protozoa, nematode etc.). At this temperature thermophilic bacteria like Thermonospora and Thermoactinomycetes and thermophilic fungi such as Thermoascus carry out these reactions.

Once the readily degradable materials are degraded the reaction rate slows down. Eventually temperature decrease to mesophilic range through heat loss from the surface of heap. Once again the mesophiles either through re-invasion from outside or through germination of heat resistant spores dominates. The thermopile process has certain advantages like acceleration of process, killing pathogenic microorganisms etc.

Mesophilic microorganisms (which digest simple organic compound)

Bacteria	***Fungi***
Cellulomonas	*Aspergillus*
Myxococus	*Penicillium*
Chondrococus	*Fusarium*
	Trichoderma

Thermophilic microorganisms (which digest cellulose, hemicelluloses, lignin etc)

Bacteria	***Fungi***
Thermonospora	*Mucor*
Thermoactinomycetes	*Aspergillus*
Nocardia	*Rhizopus*

Advantages

- Simple equipment and very little training required
- Suitable for tropical environment
- Appropriate for small cities and towns

Disadvantages

- Glass, metal etc should removed
- Farmers may refuse compost because of light soil

5.9. RECLAMATION OF WASTE WATER

Effluent from sewage treatment plants apart from discharge in to water bodies for dilution can also be reused for various purposes. Waste water reclamation has been practiced since about 1920 in United States. The first reuse of reclaimed waste water was for crop irrigation; the crops were fiber, fodder and seed crops and there was little opportunity for public contact. Later reclaimed waste water was used for landscape irrigation *i.e.*, Irrigation of parks, play ground etc. Water scarcity has lead to phenomenon of water recycling.

Industrial Reuse of Effluent

Due to the cost of treatment process, treatment and recycling of waste water will be practiced only when efficient discharge is prevented. There occurs different orders of waste water within the processing operation and depending upon the industrial process different grades of water can be used. This recycling is normally practiced in heavy industries such as for process operation and steel making where water of low quality can be used and minimum treatment of waste water is sufficient as water is only for cooling. Higher grade waste water from sterilizing and food processing units can be recycled with minimum treatment for a lower grade use such as vegetable cleaning.

Municipal Reuse of Effluents

Recycling of domestic waste water is becoming popular in arid areas. Used water from bathing, dish washers, washing machines etc can be reused for flushing toilets after treatment with mild sterilizing agents like Sodium hypochlorite etc

Agriculture Reuse of Effluent

Effluent from sewage treatment plants contain wide range of inorganic plant nutrients, thus effluent waste water can be efficiently used as a source of water and nutrients for the plants of arid climates. However there is a possibility for hazards like contamination by pathogenic micro organisms or accumulation of toxic chemicals

Irrigation

Treated waste water can be used for irrigation. It can carry out as land filtration and broad irrigation. Land filtration methods can be followed in landscapes in which are porous for filtration. This is an efficient method where effluent is irrigated in to parallel trenches which are dug with an even depth to avoid the flow of water to the lowest point. Broad irrigation method is followed when the ground is not sufficiently porous. It is less efficient than land filtration and requires the use of more land there is a main trench from where arises several side branches. Land which overdosed with sewage become sewage sick. Such lands can be restored by a prolonged rest.

Disadvantages

If the water is recycled many times, accumulation of dissolved salts particularly anions rich as nitrates, sulfates, phosphates and chlorides. This may lead to corrosion and toxicity problems.

Solid Waste Recycling

Sludge conditioning

Sludge particles contain enormous amount of bound water. To release these bound water and to allow agglomeration of solid a preliminary conditioning stage is required before dewatering process. This preliminary process is termed as sludge conditioning.

Elutriation is a method of sludge conditioning applicable for digested sludge with high alkalinity. Alkalinity in sludge exerts a high requirement for chemical conditioning. This excess alkalinity can be removed by mixing the sludge with water or effluent and allowing sludge to settle.

Sludge as Fertilizer

Considerable quantity sludge is discharged to land and can be used as a fertilizer for agricultural lands and rest is used for land fill. In certain cases the sludge is mixed with domestic reuse and treated by various methods of composting. This provides a compost containing nutrients and humans which can used to condition the soil

Advantages

Heat dried activated sludge is by far the best for use as fertilizer. It has the advantages of reduced bulk, little odor, high nitrogen content, good source of organic matter, low pathogenic bacterial content and low grease content they can be stored dry until use it is recommended as a source of nitrogen.

Disadvantages

Sludge as a fertilizer has certain disadvantages. It has a moderate manurial value and has to be prepared in such a way that its more better than a low grade manure. Only small quantities of potassium or magnesium are present in sludge and these have to be supplemented. Much of nitrogen in digested sludge is in the form of Ammonia. Sludge contain mineral, grease and potentially toxic chemicals which affect plant growth or heavy metals like Zinc, Copper, Iron, presence of pathogenic bacteria particularly *Salmonella* and eggs of parasites and make the sludge unsuitable as manure for crops.

Chapter 6

INDUSTRIAL BIOTECHNOLOGY

6.1. STRAIN IMPROVEMENT

Strain development is economic, since the metabolite concentrates (concentrations) produced by wild strains are usually too low for economic processes. Since the late 1930's the use of mutations as a method of strain improvement has been exploited in many industrial processes (most notable example being the production of Penicillin). Advantages of using improved strains are- it is cost effective strains, with improved fermentation properties, more efficient, high yielding etc. Mutants will synthesise one component as main product (instead of mixture of chemically closely related substance). rDNA technology also has immense potential and is used for the production of insulin, somatostatin, human growth hormone etc.

Mutation

The basic procedure for producing mutants from micro organisms is relatively simple and involves exposing cells or spores to the action of a mutagen. The exposure may take place on the surface of agar in petriplate or in liquid culture etc. The process usually kills 99% of exposed cells. Remaining living cells can be isolated, sub-cultured and checked for desired result.

Mutations generally are

1. ***Genome mutation:*** May cause change in chromosome number.

2. ***Chromosome mutation*:** May change order of genes within chromosome (deletion, inversion, duplication etc)

The commonly used mutagens fall into two large groups-chemical and radiation.

Chemicals

A wide variety of chemical compounds are capable of producing mutation, a transition, transversion, frame shift mutations which may result or with varying effects (some cause loss or addition of bases in DNA, change in nucleic acid etc.)

Nitrous Acid (HNO_3)

It deaminates Adenine to Hypoxanthine or Cytosine to Uracil (Hypoxanthine pairs with Cystine, Uracil with Adenine causing AT - GC and /or GC - AT transitions.

Ethyl Methane Sulphonate (EMS), Ethyl Ethane Sulphonate (EES) and Diethyl Sulphate (DES)

These groups belong to larger group known as alkylating agents and are most useful mutagenic system for practical applications, resulting in tansitions transversions and frame shift mutations.

N - Methyl N-nitrosoguanidene (NTG)

It is one of the most effective chemical agent producing high yields of mutants with high survival rate, but its use require great care because of its carcinogenic effects.

Base analogues

Compounds such as 5 Bromo uracil and 2 Amino purine have been used for mutagenic effects. These compounds are often incorporated in its DNA in place of Thymine and Adenine respectively. The mutagenic effect with base analogue are of minor importance for practical application as it can be costly method. *Eg:* Acridine compounds.

Acridine compounds can result in frame shift mutation. These frame shift mutagens can intercalate into DNA molecule and cause errors which result in alteration in reading frame resulting in a faulty protein or no protein at all. *Eg:* Acridine orange, Acriflavine etc.

Radiations

Several systems of radiations x-rays, gamma rays, uv rays etc have been used for mutagenesis. Radiation systems are relatively convenient.

Ionizing radiations

Ionizing radiations such as x-rays, uv rays, etc have been used successfully to produce mutations. This type of radiation is particularly useful whenever chemical agents or uv light is not effective. Ionizing radiations often cause chromosomal breakage which may lead to considerable structural changes.

Ultraviolet Light

The use of light between wavelength of 200 and 300 nm is particularly easy and safe way of producing mutagenesis in microorganisms.

Hybridization

The use of genetic recombination in micro organisms is valuable for producing new strains of industrial importance. Genetic recombination in fungi and yeast can be accomplished by obtaining haploid strain from ascospores, basidiospores etc. and then mating to produce new strain. The Hybridization of yeasts and higher fungi requires 3 main steps –Sporulation, Spore isolation and Hybridization.

Sporulation is controlled by many factors such as media composition, age of culture, temperature etc. Once sporulation has taken place, it is necessary to seperate asci from non sporulating cells. After asci has been separated, individual spores can be obtained by methods such us ascus dissection. Hybrids can be prepared by spore mating haploid cells etc to produce new strains.

Parasexual Cycle

Nuclear fusion and gene segregation can take place in absence of sexual organs by a process termed parasexual cycle characteristic to many industrially important fungi such us *Aspergillus niger, Penicillium chrysogenum* etc. The recombination is achieved only in an organism with at least two different types of nuclei coexist *i.e.*, a heterokaryon. The heterozygous diploid nucleus resulting from the fusion of two different nuclei give rise to diploid clone; a diploid nucleus in the clone may undergo an abnormal mitosis resulting in mitotic segregation and development of recombinant clones which may be either be diploid or haploid.

Protoplast Fusion Techniques

Protoplasts are cells devoid of their cell walls and may be prepared by subjecting cells to the action of wall degrading enzymes in isotonic solution. Cell fusion followed by nuclear fusion may occur between protoplasts of strains (which would otherwise not fuse) and the resulting fused protoplast may regenerate a cell wall and grow as normal cell. Protoplast fusion has been demonstrated in a large number of industrially important organism including *Streptomyces* spp, *Bacillus* spp, filamentous fungi and yeasts.

Genetic Recombination in Bacteria

In bacteria portion of genetic material of one cell (donor) is transferred to second cell (recipient). The recipient cell therefore becomes diploid, only in a section of etc genetic compliment. The process of recombination may proceed in one of three ways.

a. ***Transformation*:** Accomplished by adsorption of a price of double stranded DNA (which had been released into the medium by a donor cell) on to surface of a recipient cell.
b. ***Transduction*:** Which requires a bacteriophage where a portion of double stranded DNA from donor cell is carried to recipient cell by bacteriophage. When infection by bacteriophage take place genetic material from donor to recipient is incorporated to form a partial Zygote.
c. ***Conjugaton*:** Genetic material here is transferred to recipient by direct contact A single strand of DNA is transmitted.

Recombinant DNA Technique

1. A vector DNA molecule (plasmid or phage) capable of entering host cell and replicate when it is prepared so as to integrate foreign DNA (without destroying its function)
2. A method of splicing genetic information into vector is done to create a DNA
3. The vector is introduced (as DNA) into host cell and selected for their presence through characters such as drug resistance, plaque formation etc.

4. Assaying for foreign gene product of choice from the population of recombinants created.

6.2. PRODUCTION MEDIUM

The growth medium meant for cultivation of a production strain and subsequent production of either microbial cells or fermentation medium.

Characteristics of an ideal production medium are as follows:

1. ***Chemical composition***: The production medium must have a suitable chemical composition. Generally the medium should contain a source of Carbon, Nitrogen, growth factors and mineral salts and dextrins. Nitrogenous substances present include proteins, peptides, amino acids, purines, pyramidines and vitamins.
2. ***Precursors***: In certain fermentations (*Eg:* Pencillin fermentation) the medium should supply the required precursors for better yield of desirable product.
3. ***Buffering capacity***: Maintenance of PH in the optimal range is necessary for making the process successful. Since acidic and/or basic compounds depending on the nature of fomentation process accumulates during the process of fermentation, inorder to control the PH of medium, buffers (*Eg:* $CaCO_3$) should be added to the medium. Media containing considerable amounts of proteins, peptides and amino acids pocess good buffering capacity in the PH range near neutrality. Additional buffering capacity in this pH range is also provided by phosphates (*Eg:* mono and dihydrogen sodium or pottassium phosphates).
4. ***Antifoam agents***: Foaming is a serious problem in fermentation industry since it may aid in contamination and possibly cause other problems in fermentation. Hence antifoam agents (*Eg:* lard oil mixed with octadecanol for Pencillin fermentation).
5. ***Toxicity***: An ideal production medium should be free from any toxic effect of cultures.
6. ***Consistency***: In aerobic fermentation it is necessary to supply sterile air into the medium. Under such circumstances liquid media allows the diffusion of air into the medium under agitation. Fermentation medium should not be viscous, since the viscous nature of medium creates difficulty in penetration of air into the interior of the medium.

7. ***Recovery*:** Component of the medium should be such that separation and extraction of product becomes easy and cheaper.
8. ***Availability of raw material*:** The raw materials required for designing of production medium should be freely available in large quantities at reasonable price. Hence crude media (or non synthetic media) find applications on large scale production process. Synthetic media are expensive and give low yield

On large scale one must normally use sources of nutrients to create medium which meet criteria like,

1. Produce maximum yield of product per gram of substrate used
2. Permit maximum rate of product formation
3. Minimum yield of undesirable product
4. Readily available throughout the year
5. Cause minimal problems during media making and sterilization or extraction.

Raw Materials

Many different types of raw materials are used in different types of industrial fermentation process. Crude nutrition sources are preferred, since they are economical. Mostly agricultural products are utilized as source of raw materials in fermentation industries.

Media used in industrial fermentations must contain elements required for the synthesis of cell materials and for product formation. In addition media must be designed to satisfy the technical objectives of the process by providing favorable environment for growth and /or product formation, while at the same time being cost effective.

Medium Formulation

Micro organism requires Carbon, Nitrogen, minerals, sometimes growth factors, water and oxygen (if aerobic). The medium is formulated so as to create an optimally balanced medium (Typical values of C, H, O, N and S as a percentage of dry weight 45, 7, 33, 10, and 2.5 respectively). Trace elements (Cu, Mn, Co, Mo, B etc.) may be required. Precursors need to be incorporated in some cases (1AA in Penicillin production) in addition to the critical concentration of minerals.

Carbon Source

The carbon requirement under aerobic conditions may be determined from biomass yield coefficient

$$Ye = \frac{\text{biomass produced (g)}}{\text{carbon substrate utilized (g)}}$$

Care should be taken during sterilization of media containing Malt extract. When overheating occurs a Millard reaction result due to low pH and high proportion of reducing sugar. In this conversion, the amino group of amino acid or proteins react with carboxyl group of reducing sugars, aldehydes or ketones resulting in a brown condensation product which results in reduced yield.

Molasses

Molasses may be of three different types depending on the source of raw material

A. Blackstrap molasses

Component	*Percent*
Water	20
Ash	8.0
Total sugar	40.60
Total nitrogen	3.0
Total nitrogen	0.5
Gums	2.0
Free acids	2.0
Combined acids	3.0

About 95% of total sugar in cane molasses is fermentable. It is particularly rich in biotin, pantothenic acid, thiamine, phosphoric acid and sulfur. The organic Nitrogen content is less than beet molasses

Beet molasses are produced by the same process employed for cane molasses. Vitamins such us biotin, pyridoxine, thiamine, pantothenic acid and inositol are also present. The largest utilization of cane blackstrap molasses in India is in alcohol industry, which utilizes it for the manufacture of spirit (ethyl alcohol), country liquors and other liquors such as rum, brandy, gin and whisky. Cane and beet molasses are widely used in fermentation industries as shown below

Chemical composition of beet molasses

Component	***Percent***
Suerose	48.50
Raffinose	1
Glucose+Fructose	1
Nitrogen compounds	12–13
Glutamic acid	3.5
Other amino acid	5.5
(Aspargine, Aspertic acid	
Alanine, Glycine) Ash	11–12

B. ***Cane molasses***

Produced by: Bacterial action

Uses: Direct use

Latic acid (Tanning, Textile)

Acetone (Rayon, Acetic anhydride)

Ethyl alcohol (Oils and waxes, synthetic rubber, Dyes, Cellulose products)

C. ***Beet molasses***

Produced by: Fermentation by yeast

Uses: Ethyl alcohol (Oils and waxes, synthetic rubber, Dyes, Cellulose products)

Uses: Direct use, Proteins

Source obtained from sugar cane and sugar beet is commonly used in fermentation mcdia as impure form obtained as Beet or cane molasses, which are residues left after crystallization of sugar solution in sugar refining.

Molasses is used in production of high volume/low value product such us ethanol, SCP, organic acid, amino and some microbial gums. Molasses may also be used for production of higher value/low bulk products such as antibiotics, speciality enzymes vaccines and some chemicals. The cost of molasses will be very competative when compared with pure carbohydrates. However molasses contains many impurities and molasses based fermentations will often need a more expensive and complicated extraction/purification stage to remove impurities

Starch and Dextrins

Starch and dextrins can be directly metabolized as carbon source by

Amylase producing organisms. In addition glucose syrup is also used frequently in formentation.

Sulfate waste liquors

Sugar containing waste products of paper industry which have a dry weight of 9–13% are primarily used for cultivation of yeasts.

Cellulose

Exist as waste in forms as straw, corn, wood wastes, peat waste, paper etc. The World annual cellulose production is estimated to be 10 metric tons. It is not usually possible to use cellulose directly, instead must be hydrolyzed to sugar syrup (by chemical or enzymatic treatment). This can be used for ethanol fermentation, acetone and isopropanol production.

Whey

Whey is a byproduct of dairy industry. It is produced annually on Worldwide basis to amount of 74 million tons containing 1.2 million tons of lactose and 0.2 million tons of milk protein. The lactose is used primarily for production of ethanol or SCP, Xanthan gum, Vitamin B12, Lactic acid and Gibberellic acid

Animal fats such as lard and animal and plant oils are readily utilized by some micro organism but are generally added as supplemental substrates rather than sole source of carbon.

Methanol is cheapest fermentation substrate with respect to content but can be metabolized only by a few bacteria and yeasts (even for SCP), ethanol available in ample supply from fermentation of saccharified starch or cellulose can be metabolized by many micro organisms.

Substrates as Nitrogen Source

Many large scale processes utilizes ammonium salt, gaseous ammonia etc. as Nitrogen source. But industrial production phases in production of other industrial products may use wide variety of substances available on mass scale from various nitrogenous substrates.

Yeast Extract

They are produced from baker's yeast through autolysis at 50–55°C or through plasmolysis in presence of high concentration of NaCl. Yeast extract contain amino acid and peptides, water, vitamins and carbohydrates.

Peptone (Protien hydrolysates)

Can be utilized by many micro organisms, but is expensive for industrial applications. Source of peptone include meat, casein, gelatin, keratin, peanut seeds, soy meal, cotton seeds, sunflower seeds etc. Peptone combination varies depending upon origin (*Eg:* peptone from gelatin is rich in proline and hydroxyproline ; peptone from plant origin like soy peptone has large proportion of carbohydrates).

Soymeal

Soy meal is residue from soybeans after extraction of soybean oil; is a complex substrate with 50% protein, 30% carbohydrate (1% fat, 1.8% lecithins). Soy meal is used in Antibiotic fermentations. Catabolite regulation does not occur because of slow catabolism of this complex mixture.

Complex Substrate

Crude complex substrates provide a cheap source of carbon, nitrogen and other nutrients. They include whole plant tissues, animal and microbial by p

6.3. RAW MATERIALS FOR PROCESS CONTROL

Fermentation raw materials for process control include

1. Buffers (*Eg:* inorganic phosphate added for a buffering in ranges pH 6.0 to 7.5)
2. Acids (Sulfuric acid, Hydrochloric etc)
3. Alkalies ($CaCO_3$)
4. Antifoam agents

Antifoams

Foam frequently occurs in fermentation due to the denaturation of proteins at gas liquid interface. If uncontrolled, foam may rise to fill the fermentation head space and can eventually result in evacuation of much of fermentation contents through air exhaust outlet. Foaming can also cause removal of cells from medium.

Antifoams are surface active agents which act by reducing surface tension in foam to disperse foams

Ideal antifoam should have the following properties.

1. It should readily disperse and have fast action in existing foam
2. It should be active at law concentration and should be long lasting
3. It should be non toxic and should not cause problem in downstream processing.
4. It should be non toxic and non hazardous and should be cheap (*Eg:* Alcohol, fatty acid etc.)

Animal Cell Culture

Sera widely used include fetal calf serum, new born calf serum and horse serum. Serum contains several factors essential for cellular competence and proliferation including initiation factors, binding proteins and low molecular weight compounds.

Serum free media has been developed which contain carbohydrate, amino acid, vitamins, nucleic acid etc. dissolved in high purity water. This makes reduction in potential contamination to make sterility easier to achieve and potential cost savings because of cheaper replacement components.

Nutrient Media for Plant Cell Culture

They have chemically defined media containing organic carbon source, nitrogen source, inorganic salts and growth regulators. Plant hormones like Auxins and cytokines may also be required.

Buffers

The control of pH may be extremely important if optimal productivity is to be achieved. Many media are buffered at about pH 7.0 by the incorporation of Calium carbonate (as chalk). Phosphate which are part of many media also play an important role in buffering. The balanced use of carbon and nitrogen source will also form a basis for pH control as buffering capacity can be provided by proteins, peptides and amino acids such as in corn steep liquor. The pH may also be controlled externally by addition of Ammonia or Sodium hydroxide and Sulfuric acid.

Precursors

Some chemicals when added to certain fermentation are directly incorporated in to desired product. *Eg:* A range of different side chains can be incorporated into Penicillin molecule.

Examples of precursors used on fermentation processes

	Precursor	*Product*	*Micro organism*
1.	Phenyl acetic acid	Penicillin G	*Penicillium chrysogenum*
2.	Phenoxy acetic acid	Penicillin V	*Penicillium chrysogenum*
3.	Chloride	Griseofulvin	*Penicillium griseofulvin*
4.	Cyanide	Vitamin B12	*Propionibacterium*
5.	α-amino butyric acid	L-Isoleucine	*Bacillus subtilis*

Inhibitors

When certain inhibitors are added to fermentations, more or a specific product may be produced or a metabolic intermediate which is normally metabolized is accumulated. In most cases the inhibiters is effective in increasing the yield of desired product and reducing the yield of undesirable related product. Inhibitors have also been added to affect the cell wall structure and increase the permeability for release of metabolites (*Eg:* use of *Penicillium* and surfactants in glutamic acid production)

Product	*Inhibitor*	*Effect*	*Micro organism*
Glycerol	Sodium bisulphate	Acetaldehyde production repressed	*Saccharomyces*
Tetracyclin	Bromide	Chlorotetracycline repressed	*Streptomyces*
Glutamate	Penicillin	Cell wall permeability	*Micrococcus*
Citric acid	Metal phosphate	Oxalic acid repressed pH below 2	*Aspergillus niger*

Foaming due to component in medium or some factor produced by micro organism is a serious problem in industries. The most common cause of foaming is due to proteins in medium such as corn steep liquor, yeast extract, soybeen meal. These proteins may denature at air broth interface and form a skin which does not readily rupture. Foaming can cause removal of cells from medium which will lead to autolysis, further release of microbial cell proteins etc. This result in problems including reduction in working volume of fermentor due to oxygen exhausted gas bubbles circulating in the system. The biological problems include deposition of microbial cells in upper parts of fermentor, loss of product etc.

Antifoams are surface active agents, reducing the surface tension in the foam and destabilizing protein films by

a. Hydrophobic bridges between two surfaces
b. Displacement of absorbed protein
c. Rapid spreading on the surface of the film.

Ideal antifoam should have the following properties.

1. Should disperse readily and have fast action on existing foam
2. Should be active at law concentrations.
3. Should be long acting preventing new foam formation
4. Should not be metabolized by micro organism
5. Should be non toxic to micro organism, humans and animals.
6. Should not cause any problems in extraction and purification of the products.
7. Should be cheap
8. Should be heat sterilizable etc.
9. Should have no effect on Oxygen transfer.

   Egs:-

   - *Alcohols*: Stearyl and octyl decanol
   - Esters
   - Falty acids and derivatives particularly glycerides (cotton seed oil, linseed oil, soy bean oil, olive oil, castor oil, sun flower oil, cod liver oil etc.)
   - Sulphonates
   - *Miscellaneous*: Polypropylene glycol

In addition to this, growth factors are required in some instance, some micro organism cannot synthesize a full complement of cell components and require growth factors *Eg:* Amino acids, vitamins etc.

6.4. DESIGN OF FERMENTOR

Fermentors provide a controlled environment for the growth of microorganisms or animal cells to obtain a desired product. They are complicated in design, since they must provide for the control and observation of many stages of microbial growth and biosynthesis. The design of the fermentor depends upon the purpose for which it is to be utilized.

In designing and constructing a fermentor, a number of points must be considered

1. The vessel should be capable of being operated aseptically for a number of days and should be reliable in long -term operation
2. For aerobic submerged fermentation the tank should be equipped with aerating devise
3. There should be stirring device for uniform distribution of air nutrients and micro organisms
4. A fermentor should permit easy control of contaminating microbes.
5. A fermentor should be provided with the inoculum point for easy transfer of inoculum
6. Baffles should be present to avoid vortex formation.
7. Sampling valves should be present with the fermentor for withdrawing the sample during the process for different laboratory tests.
8. There should be a provision for controlling the pH and the temperature of the medium
9. There should be a facility for intermittent addition of an Antifoam agent
10. A drain at the bottom is essential for the removal of the completed fermentation broth for further processing.
11. A manhole should be provided at the top for access inside the fermentor for different purpose like repairing and cleaning.

Fermentors are available in varying sizes. These sizes are usually stated based on the total volume of capacity of the fermentor.

Head Space

The actual oprating volume in a fermenter is always less than that of total volume, because a space must be left at the top of fermentor to allow splashing and also for complete aeration and agitation. This space is known as head space.

pH Control

During the fermentation process due to the accumulation of products, the pH of fermenting liquid may vary. It will reduce the growth rate or activities of micro organism. In order to control the changes in pH acid or alkali is added. The addition is controlled by an Autotitrator. The autotitrator is connected with a pH probe. The pH probe helps in checking of the pH.

Temperature Control

Heat will be produced by microbial activity and mechanical agitation and the heat generated by these two process is not ideal for particular manufacturing process than heat may have to be added or removed from the system. Temperature control is achieved by water jacket around the vessel. This is often supplemented by the use of internal coils in order to provide sufficient heat transfer surface. The sterilization can also be done through this kind of coils. Hot water or cooled water is supplied trough the coil for temperature control.

Agitation

The agitating device consists of a strong and straight shaft to which impellers are fitted. An impeller consist of a circular disc to which blades are fitted with bolts. Different types of blades are available and are used according to requirements. It is rotated with the help of motor and the speed also can be adjusted. It is also possible to adjust the height of impeller blade from the bottom of the fermentor according to ones desire.

Agitators may be classified as disc turbines and vaned discs

1. ***Disc turbines***: Consist of disc with a series of rectangular vanes set in vertical plane around the circumference and the vaned disc has a series of rectangular vanes attached vertically to the underside.
2. The vanes of variable pitch open turbine and blades of marine propeller are attached directly to the bors on agitator shaft. In this case the air bubbles do not initially hit the surface before dispersion by vanes.

Baffles

The liquid medium is thrown up towards the walls of the fermentor while rotating the impellor blades of a greater speed. The result is the formation of vertex which is eliminated usually by four equally spaced baffles attached to the walls of the fermentor.

Aeration

A sparger used for aeration may be defined as a device for introducing air in to liquid or a fermentor. An aerating device consist of a pipe with minute holes through which pressurized air escapes into aqueous medium in the form of tiny air bubbles. The aeration device is basically of three types- porous sparger, the orifice sparger (a perforated pipe). A combined sparger agitator may be used in laboratory fermentors.

The size of a hole in sparger ranges from 1/64 to 1/32 of an inch or larger. A hole smaller than this requires high air pressure for an economical air bubble formation. But smaller the air bubbles greater will be the bubble surface area, so it is desirable to adjust the size of air bubble to give the greatest possible aeration without greatly increasing the overall cost of fermentation process. As a result the oriface now face downwards and fermentations run sucessfully.

These are various ways for introducing air in to fermentation vessel. The principal modes of injecting air are

1. ***Impeller air injection*:** Air is fed in to impeller by means of hollow salt injected in to the medium through hole in the impeller.
2. ***Two phase injection*:** A two phase mixture of air and nutrient medium is fed in to the fermentor in the form of foam or suspension
3. ***Air lift fermentor*:** Air is used to circulate the contents of the fermentor either through tubes external to the vessel or internally by using draft tube.
4. ***Sparger air injection*:** The most widely used fermentor is stirred baffled upright cylinder with air sparger.

Time

The time required for the fermentation process varies from hours to weeks depending upon the conversion being attempted.

Commonly Used Spargers

***Oriface sparger*:** In small stirred fermentors, the perforated pipes were arranged below impeller in the form of rings (ring sperger).

***Nozzle sparger*:** Single, open or partially closed pipe as sparger to provide stream of air bubbles, positioned centrally below impeller.

***Combined sparger agitator*:** For small scale, where air is introduced via a hollow agitator shaft and emitting it through holes drilled in the disc between the blades and connected to the main shaft.

6.5. TYPES OF FERMENTATION

Surface Process

Surface process can be used for the production of compounds like organic acids *Eg:* citric acid is produced by surface and by submerged culture process. Surface process can be further subdivided according to the state of the culture medium used.

Solid State Fermentation

Surface process employing solid substrates may use either wheat bran or pulp from sweet potato starch etc as culture medium. After sterilization, the material is inoculated with spores, spread on trays in layers 3.5 cm thick and incubated at 28°C. The solid substrate process may take up to 90 hours, at the end of which the entire solution is extracted with hot water to isolate the organic acid (citric acid). It is necessary to supply air to the surface of the seeded medium.

The recovery of product (citric acid) is practiced as: The fermenation liquor obtained after treatment with hot water is precipitated, filtered, washed and crystallized.

Liquid State or Surface Fermentation Method

Surface process using liquid nutrient solution is oldest production method and account 20% Worlds supply of citric acid. These process are still in use because of low investment, low energy cost for cooling system and simple technology. In modern systems continuous sterilization of medium is used. The sterilized nutrient solution automatically flows over distribution system on to the trays. The production medium is usually kept in shallow pans on such a way that shallow layers of medium with the depth of 1–2.5 cm (but sometimes upto 8 cm) is formed. Inoculation is done in incubation chamber usually by blowing dry spores or spraying spore suspensions. Air currents and proper ventilation is provided within 24 hours after inoculation. Germinating spore form a thin cover of mycelium floats as thick convoluted white layer on the nutrient solution. The fully developed mycelium floats as thick convoluted white layer on nutrient solution. The fermentation is stopped after 8–14 days. During recovery of product (1.2–1.5 kg citric acid monohydrate/m of fermentation surface /hour for citric acid)mycelium and nutrient solution are removed from chamber, purified and crystallized.

Submerged Culture Process

Three factors are important for production in submerged process

a) Quality of material used to construct the fermentor
b) Mycelial structure
c) Oxygen supply

In this process the fungus is grown dispersed through a liquid production medium. Usually fermenation is carried out on fermenation vessel consisting of sterilizable tank with capacity of thousands of gallons. The fermenation tank is equipped with mechanical agitation and aeration.

In practice the spores (as in citric acid production) are introduced under controlled aseptic conditions. When harvested they are used in specific quantities to seed the inoculum tank containing a medium designed to develop cellular mass rather than to produce the compound. Then it is transferred to the production tank under aseptic conditions. The production media consist of constituents to product formation rather than growth. Proper provision for aeration and agitation is required, with provision to withdraw samples at regular intervals.

Submerged Culture Process for Citric Acid Production

The structure of mycelium that forms in submerged culture during Tropophase is vital to successful production process. If mycelium is loose and filamentous, with limited branches and no chlamydospores, little citric acid is produced in idiophase. Myclium for optimal production rates consist of very small solid pellets, with aeration rate of 0.2–1 volume/volume/minute (vvm). Most fermentors for citric acid production are constructed in a range of 120–220 m. Submerged system are less sensitive than surface system to variability in the medium composition and, they can be operated much more reproducibly than surface systems.

Fermentation may be carried out as batch, continuous and fed batch processes. The mode of operation is to a large extend, dictated by the types of product being produced.

Batch Culture

Batch culture is a closed culture system which contains an initial, limited amount of nutrient. The inoculated culture will pass through a number of phases.

Growth of typical microbial culture in batch condition

After inoculation there is a period during which it appears that no growth take place; this period is referred to as lag phase and may be considered as a time of adaption. In a commercial process the length of lag phase should be reduced as much as possible, this is achieved by a suitable inoculum.

Following a period during which growth rate of cells increases, cells grow at constant maximum rate and this period is known as log or exponential phase.

The exponential phase may be discribed by the equation

$$dx/df = ux$$

where x is the concentration of microbial biomass

t is the time in hours

u is the spesific growth rate in hours

Exponential growth phase is seen in bacteria's which reproduce by binary fission, indeed animal and plant cells in suspension grows exponentially. Mycelial organisms which show apical growth also grow exponentially.

Growth results in the consumption of nutrients and excretion of microbial products, and after a certain time the growth rate of culture decreases until growth cases. The cessation of growth may be due to deletion of essential nutrients in medium or due to accumulation of some toxic product of organism in the medium. The growth rate will not be affected until the substrate concentration has declined to a very low level. The deceleration phase for such culture would be short.

The stationary phase in batch culture is that point where growth rate has declined to zero (population is still metabolically active and may produce product such as secondary metaboletes at this phase).

Batch fermentation may be used to produce biomass production, primary metabolites and secondary metabolites. For biomass production, cultural conditions supporting growth rate and maximum cell population would be used, for primary metabolite productions extended exponential phase is used and for secondary metabolite production short exponential phase with extended production phase must be used.

Continuous Culture

Exponential growth in batch culture may be prolonged by the addition of fresh medium to the vessel. Exponential growth will produce until the additional substrate is exhausted; however when an over flow device was fitted to fermentor such that the added medium displaced an equal volume of culture from the vessel, then continuous production of cells could be achieved. If the medium is fed continuously to such a culture at a suitable rate a steady state is achieved eventually, *i.e.*, the formation of new biomass by culture is balanced by the loss of cells from the vessel.

The flow of medium into vessel is related to the volume of the vessel by the term dilution rate

$$D = F/V$$

where F is the follow rate (dm^3h)

V volume (dm^3)

D is expressed in units h^{-1}

The growth of cells in continuous culture is controlled by the availability of the growth limiting chemicals component of medium - system is described as Chemostat.

An alternate type of continuous culture to chemostat is the Turbidostat, where the concentration of cells in the culture is kept constant by controlling the flow of medium such that turbidity of the culture is kept within certain narrow limits. This may be achieved by monitoring the biomass with photoelectric cells (these feed signal to pump supplying medium to culture; the pump is switched on if biomass exceeds the set point and switched off if biomass falls below the set point)

The chemostat is more commonly used system became it has advantage over biostat of not requiring complex control system to maintain steady state. The basic chemostat may be modified in a number of ways, most common being addition of extra stages.

In multistage chemostat different conditions prevail in separate stages. This may be advantages in utilization of multiple carbon sources and in production of secondary metabolites (*Eg:* use of glucose in the first stage and maltose in next stage).

6.6. DOWN STREAM PROCESSING

The extraction and purification of fermentation products are known as downstream process. The recovery of microbial product may vary from as low as 15% to as high as 70% of total manufacturing cost. After fermentation the medium contains the specific product, intact microbial cells, cell fragments, soluble and insoluble medium components and other metabolic products. The product may be heat labile, intracellular and easily degradable by external factors and contaminating micro organisms. All these fact make the recovery process a difficult one. The time of recovery process depends on the following criteria

1. The intracellular or extracellular location of product
2. Concentration of product in the medium
3. Physical and chemical properties of the product
4. The minimum acceptable standard of purity
5. Other impurities in the medium
6. The marketable prize of the product etc.

Stages of Downstream Processing (DSP)

The main objective of first stage of recovery of an extracellular product is

1. Removal of large solid particles and cells usually by centrifugation or filtration.
2. In the second stage the broth is fractionated or extracted in to major fractions using ultra filtrations reverse osmosis, adsorption chromatography, liquid extraction or precipitation
3. In the third stage the product containing fraction is purified by fractional precipitation and further more precise chromatographic techniques.

4. In the last step crystallization is done.

To simplify all these processes another method is practiced known as "whole both" processing which involves initial removal of large particles which is then followed by passage of broth (including cells through ion exchange columns or counter current solvent extraction to extract the product directly.

Removal of Microbial Cells or Other Solid Matter

Most commonly used methods are filtration. Other methods are foam separation and precipitation. Filtration is one of the most common processes used in downstream process. It shows a porous medium which retains the particles but allows gas or liquid to pass through it. The factors that are to be considered in filtration process are

1. The properties of the filtrate *Eg:* Viscosity and density.
2. The size and shape of solid particles to be separated.
3. The scale of operation.
4. The need for aseptic conditions.

As the filtration precedes the filter cake gradually builds up. Due to this the resistance against the flow will increase. So the pressure applied on slurry should be increased slowly in order to keep the rate of flow of filter a constant. The flow rate may also be reduced by blocking of holes in filter cloth when particles are compressible and soft.

Use of filter aids

It is common practice to use filter acids while filtering bacteria or other fine gelatinous suspension. Kieselguhr (Diatomaceous earth) is the mostly used material. It is mixed with initial cell suspension to improve priority of filter cake and result on faster flow rate. Hence it prevents wedging of particles which are quite large enough to pass through the pores known as "blinding". Filter aids cannot be used when product is extracellular and it will absorb some of filtrate.

We can classify the filter broadly into

a. ***Batch Filter*:** Pressure leaf filter, plate and frame filters
b. ***Continuous Filter*:** Cross flow filters.

Centrifugation

When filtration is found to be unsatisfactory centrifugation is used. Filtration has the following disadvantages.

1. Filtration is slow and difficult
2. Cells or other suspended matter must be obtained free of filter aids.
3. The rate of sedimentation of particles is directly proportional to the square of diameter of particles.

Flocculation

The use of flocculating agents enhances removal of microbial cells and suspended colloidal matter. The cells will aggregate and diameter is hence increased. It will result in increasing sedimentation rate.

The flocculating agent is added with the fluid in stirred tank. It is called coagulation, which is followed by gentle agitation. During this period small flocs will be formed, which will again aggregate to form larger flocs. The common flocculating agents used are alum, quaternary ammonium compounds, tanic acids etc.

[The examples of centrifuges used in large scale industries are basket centrifuge, multichamber centrifuge, disc bowl centrifuge etc].

Cell disruption

Physico-chemical methods used are

a. Liquid shear
b. Solid Shear
c. Agitation with abrasive
d. Freeze thawing
e. Ultrasonication

Liquid Shear

It is used in large scale enzyme recovery. In this method microbial slurry is passed through non return valve and impinges against the operating valve set at a selected operating pressure. The cells then pass through a narrow channel between the valve and impact by sudden pressure drop at exit to narrow orifice cause shock and results disruption of cell.

Solid shear

Pressure exclusion of frozen microbes at around –25°C through a small oriface is practiced in this method.

Agitation with abrasives

Mechanical cell disruption can also be achieved in a disintegrator containing a series of rotating discs and small beads. The beads are made up of mechanically resistant materials such as glass, alumina, ceramics etc.

Freeze Thawing

Freezing of material cells paste will lead to formation of ice crystals and thawing will results in expression of crystals which lead to some subsequent disruption of cell. It is slow with limited release of cellular material. So it is often practical with combination of other methods.

Ultrasonication

High frequency vibrations (approximately 20 KHZ) at the tip of an ultrasonication probe leads to shock waves which causes cell disruption. For large scale use it is not found to be satisfactory due to high power requirement, large heating effect etc.

Chemical Methods

Detergents

Detergents damage the lipoproteins of cell membrane which leads to release of intracellular components. *Eg:* Sodium lauryl sulphate, Sodium dodecyl sulphate, Triton x 100

Osmotic Shock

Osmotic shock caused by sudden change in salt concentration will cause cell disruption. *Eg:* Extraction of luciferase enzyme from *Photobacterium fischeri*.

Alkali treatment

This might be used for hydrolysis of microbial cell wall material only if desired enzyme will tolerate a pH of 11.5 to 12.5 for 20–30 min.

Enzyme treatment

Some enzyme specifically act upon cell wall. *Eg:* Lysozyme

But remains as a very expensive methods

Liquid - liquid extractor

The separation of a component from a liquid mixture by treatment with a solvent in which particular desired component is preferentially stable is known as liquid-liquid extraction. The specific requirement is that a high % extraction of a product must be obtained but concentrated in a small volume of salt. The polar liquid used mixes and dissolves the salts and other polar compounds. The non polar compounds are dissolved in non polar solvents. The solvent is recovered back from caffinate (without product)

Two Phase aqueous extraction

This system has high water content and is regarded as biocompatible. Two phase aqueous systems, have hydrophilic polymers to create two aqueous phases. When hydrophilic polymers are added to aqueous solution two phases appears. *Eg:* Propyl alcohol/dextrin. The distribution of solute between two phases is characterized by partition coefficient.

Foam Separation

Foam seperation by addition of certain foam forming agents as quaternary ammonium compounds works on difference in surface tension created by surfactants (Collectors) where foam formed on surface with microbial cells can be removed.

Chromatography

This technique is used in fermentation process to isolate and purify relatively low concentration of metabolic product. Hence when liquid is passed through a column the different solutes are separated.

Adsorption Chromatography

It involves bending of solute to solid phase primarily by weak Vanderwaals forces. The material used for this purpose to pack columns includes inorganic and organic adsorbants. *Eg:* Aluminium oxide, Silica gel, Aluminium hydroxide

Ion exchange chromatography

It is defined as reversible exchange of ions; it may be liquid a solid phase used which are called *ion exchange resins*. They are of two types:

a. *Cationic ion exchange resins*: Normally contains a Sulphonic acid or Carboxylic acid active group. Positively charged solutes will bind to the resins, later they are washed off by buffers (*Eg:* Carboxy methyl cellulose).

   Eg: Streptomycin recovered with weak acid cationic resin such as Amberlite 1 Kl – 150

b) *Anionic ion exchange resins*: Normally contains anionic active group and bind negatively charged solute to the resins. Later they are washed off by buffers.

Both cationic and anionic exchange resins are used.

Affinity Chromatography

The mechanism is similar to Adsorption chematography. It is used for purification of biological molecule on the basis of chemical structure. It depends on highly specific interactions between Antigen-Antibody, Enzyme-Substrate, Enzyme-Inhibitors etc.

The molecule to be purified is specially absorbed from a mixture. The absorbing column contains binding substances called ligand and solid support called matrix (The ligand is attached to matrix by physical absorption or chemically by covalent bond).

HPLC (High Performance Liquid Chromatography)

Improvements in nature of column packing materials yield more rigid and more uniform beads. It allows packing in column with maximum spaces between beads. Since high pressure is used to drive solvents through silica gels packed beads it is called HPLC.

Ultra Filtration

It is a process in which solutes of high molecular weight are retaining when solvent and low molecular weight solutions are forced to pass through membrane of very fine pore size.

Reverse Osmosis

Solvent molecules are forced by an applied force to move through semi-permeable membrane in a mode of ion displacement. It is used to smaller molecules.

Drying

Drying of any product is often last stage of manufacturing process. It involves final removal of water from each sensitive material ensuring that there is minimum loss of viability, activity or nutritional value.

Advantages

1. Most of transport can be reduced
2. Dried material is easy to handle and storage is more convenient
3. It is possible to remove as much as water initially by centrifugation or in filter press to minimize cost in drying process. The most common types of dryers in use are:
   a. *Contact dryers*: In contact dryers the product is in contact with hot surface *Eg:* Drum drier. It has two slowly rotating steam heated drum on to which the slurry is passed. Evaporation take place and dry product is removed slowly with the help of scraper plate.
   b. *Spray dryers*: It is used when starting material is in the form of liquid. Once material come in contact with hot surface, the material is dried by spraying through a narrow nozzle to obtain small droplets which fall in to spiral steam of hot gas at 150–250°C. The high surface area by boiling helps in rapid evaporation and complete drying within a few seconds.

   Another method is freeze drying which is used in preservation and storage of biologicals and pharmaceuticals. The material is frozen first and dried in vacuum.

Crystallization

It is used in initial recovery and also in final purification of many compounds. Bacically it is a chemical reaction in which crystals are formed which can be collected from solution. *Eg:* In citric acid production filtration broth is treated with $Ca(OH)_2$, so that relatively insoluble Ca filtrate will be precipitated in column or Pottassium acetate for Penicillin G.

Whole Broth Processing

It is a concept of recovering a metabolite directly from an unfiltered fermentation broth. It is a simple and direct process.

Streptomycin is directly absorbed on to a series of cationic exchange resins columns

Novobiocin is harvested and broth is filtered first through screen to remove larger particles fed into series of resin columns.

Stock Cultures

Microorganisms that carry out fermentations are of value if they are stored for future use. The maintenance of such cultures form stock cultures with their growth and production capacities unaltered. Maintenance of stock cultures plays an important role in industrial fermentation, research and production.

There are two types of stock cultures:- *Working stocks* and *Primary stocks*. Working stock cultures are used frequently and must be maintained in vigorous and uncontaminated condition. These cultures are maintained as agar slants; agar stabs, spore preparation of both culture and are held under refrigeration. They must be checked constantly for possible changes in growth characteristics nutrition productivity and contamination.

Primary stock cultures are cultures that are held in reserve for presently practical or new fermentation for biological assays or for possible later screening programs. These culture are not maintained in a state of high physiological activity and they are delved into only rarely. Transfer from these cultures is made only when a new working stock culture is required or when primary stock culture must be subcultured to avoid death of cells. Primary stock cultures are stored in a manner to require the least possible numbers of transfer a period of time. Primary stock cultures stored at room temperature are maintained in sterile soil or in a gas or both that is provided with an overlay of sterile mineral oil. Agar and broth cultures are also maintained in low temperature. Finally primary stock cultures are lyophilized or freeze dried and stored at low temperature.

Chapter 7

ELECTRON MICROSCOPY

ELECTRON MICROSCOPY

In electron microscope, a beam of electrons is used as a source of illumination. Electrons are negatively charged subatomic particles having a mass of 9.1×10^{28} grams. When a metal lake tungsten is heated by applying a high voltage current, electrons came out in a continuous stream and this can be directed to form a high velocity electron beam. The electron microscopes have basic constructions pattern with

1. An electron generator as sources of illumination, also called *electron gun* has a tungsten filament which when heated by electric current emits stream of electrons. The electrons accelerated by high voltage energy are formed through a collimating apparatus which renders the rays in parallel lines and fashions them into a beam. From there, the beam passes though condenser. The condenser lens is a magnetic coil which corrects the aberrations (binding) in the beam and directs them towards the object.
2. Specimen supporting grid is unusually a colloidal film or polymerized plastic plus a screen grid of copper. The specimen kept below the condenser and electron beam passes through it is scattered depending on varying refractive index of specimen. The electron beam coming out of specimen passes through second set of magnetic coils- objective lens which focus the electrons and form an intermediate image. A third set of magnetic lens -the projection lens produce the final image.

3. The final image cannot be viewed as rays of electrons are harmful to eye. They are either projected on fluorescent screen or allowed to expose photographic plate.
4. The entire electron microscope must be in vacuum or otherwise electrons get scattered due to collision with air molecules and fail to get forwarded.

Transmission Electron Microscope (TEM)

TEM has a practical resolution roughly 1000 items shorter than that of visible light. TEM has a practical resolution roughly 1000 times better than light microscope (Points closer than 5A° or 0.5 nm can be distinguished.

A heated tungsten filament in electron gun generates beam of electrons that is then focused on the specimen by the condenser. The specimen scatters electrons passing through it and the beam is focused by magnetic lenses to form an enlarged visible image of specimen on fluorescent screen. A denser region in specimen scatters more electrons and therefore appears darker in image, since fewer electrons strike that area of screen, in contrast electron transparent regions are brighter. The screen can also be moved aside and image captured on photographic film as permanent record.

Since electrons are quite easily absorbed and restored by solid matter, only extremely thin slices of microbial specimen can be viewed in TEM (20–100 nm thick). Such a thin slice cannot be cut unless the specimen has support of some kind, the necessary support is provided by plastic. After fixation with chemicals like gluteraldehyde or osmium tetroxide to stabilze all cell structure, the specimen is dehydrated with organic solvents (e.g., Acetone or ethanol)

Complete dehydration is essential because plastics used for embedding are not water soluble. Next the specimen is soaked in unpolymerized liquid epoxy plastic until it is completely permeated and plastic is hardened to form a solid block. Thin sections are cut from this block with glass ultra microtome.

Specimens are prepared for observations by soaking their sections with solutions of heavy metals like Lead citrate and Uranyl acetate. The lead and uranium ions bind to cell structures and make them more electron opaque thus increasing contrast in material.

Although specimens usually are embedded in plastic, other useful techniques include negative staining, shadowing with metal etc.

TEM will also disclose the shape of organelles within microorganisms, if specimen are prepared by freeze etching procedure. Cells are rapidly foreign in liquid nitrogen and warmed to –100°C in vacuum chamber. Next a knife that has been pre-cooled with liquid Nitrogen (–196°C) fractures the foreign cells which are brittle and break along lines of greatest weakness. The specimen is left in high vacuum for a minimum or more so that some ice can sublime away and uncovers more structural detail. Finally the exposed surfaces are shadowed and coated with layers of platinum and carbon to form replica of the surface. Replica when studied in TEM to provide a detailed three dimensional view of intracellular structures.

Scanning Electron Microscope

Scanning electron microscope (SEM) has been used to examine the surface of micro organisms in great detail. The SEM differs from other microscope in producing the image from electrons emitted by objects surface rather than from transmitted electrons.

Micro organisms must be fixed dehydrated and dried to present surface structure and prevent collapse of cells when they are exposed to SEM's high vacuum. Before viewing, dried samples are mounded and coated with thin layer of metal to prevent the buildup of an electrical charge on the surface and give a better image.

SEM scans a narrow tapered electron beam back and forth over the specimen. When the beam strikes a particular area, surface atoms discharge a tiny shower of electrons called secondary electrons and these are trapped by special detector. Secondary electrons entering the detector strike a scintillator causing it to emit light flashes that a photomultiplier converts to an electrical current and amplifies. The signal is sent to cathode ray tube and produces an image like a television picture which can be viewed or photographed.

The number of secondary electrons reaching the detector depends on nature of specimen surface. When electron beam strikes raised area, large number of secondary electrons enters the detector. In contrast fewer electrons escape a depression in surface and reach the detector. Thus raised areas appear lighter on the surface and depressions darker. Thus a realistic three dimensional image of microorganism's surface with great depth of focus results.

Chapter 8

CONTROL OF MICRO ORGANISMS

Laminar Air Flow Chamber

Laminar air flow hood is used for reducing the danger of infection while working with infectious micro organisms for preventing contamination of laboratory materials. It works on the principle of application of High Efficiency Particulate Air (HEPA) filters, which remove 99.97% of 0.3 cm particles. Laminar flow biological safety cabinets force air through HEPA fitters which provide vertical curtain of sterile air across the cabinet providing air free of microorganisms.

Filtration

Filtration is used extensively to sterilize liquid or gas including human or animal culture media, serum, enzymes, toxins, antibiotics etc, which can be easily destroyed by heat. A number of bacteriological and serological filterer are available for this purpose

Porcelain filter (Chamber land filter) is composed of hydrous Aluminium silicate or kalonin and are baked at a temperature as high as possible without sintering the clay. The filters have various porosity ranging from 0.65 to 15 µm. They are cleaned by heating to burn off organic matter in the pores.

Diatomaceous Filters

Diatomaceous earth is fine siliceous powder composed of diatoms (also called Kieselguhr). It is prepared by mixing diatomaceous earth, asbestos and organic matter such as plastic of pores in different proportions depending upon the size of the pores. The mixture is subjected to high pressure and baked of 1000–2000°C to bind materials together. They are also called Berkefield filters and can be cleaned by washing.

Fritted Glass Filters

There are formed by heating powdered glass in a disk form to a temp just below its melting point. The disk is then fused in glass funnel and cleaned by treatment with Sulfuric acid containing sodium nitrate. The strong acid weekly oxidizes and dissolves organic matters and dissolves organic matter. The acid is removed through washing.

Asbestos Filters

Are compared of matted asbestos fibers in disk foam and are called seitz filters. The disk is clamped tightly between two smooth metal rams by screw clamps. Bacteria and other suspended particles are adsorbed by fitter disks. These desks are discarded after use.

Membrane Filters

These are composed of cellulose and are called polypore or Millipore filters. The filters are discarded after use. It is kept within membrane filter assembly and negative pressure is applied by use of vacuum to force fluid through filters. Efficiency of filter is also due to filtration controlled by electric charges on the filter. Wide variety of pore sizes are available (pores of about 0.2 µm are widely used)

Radiation

Radiation as ionizing and non ionizing radiations are used for sterilization

Ultraviolet (UV) radiation around 260 mm is quite lethal but does not penetrate glass, water, dirt and other substances very effectively. Hence it is used only in few situations. UV lamps are sometimes placed on wings of rooms or in biological safety cabinets to sterilize the air and any exposed surface. Commercial UV units are available for water treatment because uv radiation irritates eyes and damages skin and eyes. People working in such areas must be certain the uv lamps are off when the areas are in use.

Ionizing radiation is an excellent sterilizing agent and penetrates deep to objects. It will destroy bacterial endospores and vegetative cells; however it is not effective against viruses. Gamma radiation from cobalt to is used in cold sterilization of antibiotics, hormones, plastic disposable supplies such as syringes etc. It is also used to sterilize meat and other food. Irradiation can eliminate the threat of *E. coli Staphylococcus aureus* etc

UV rays form Thymine dimer (occasionally cytosine dimers) which interferes with replication and transcription whereas ionizing radiations destroy microbes through ioinization and formation of free radicals.

Chemical Methods of Sterilization

Phenol

Phenol was first widely used antiseptic and disinfectant. In 1867 Joseph Lister employed it to reduce the risk of infections during operations. Phenol and phenolics (Phenol derivatives such as cresols xylenols, ortho phenylphenol) are used as disinfectant in laboratories and hospitals. The commercial disinfectant lysol is made of mixture of phenolics. They are effective in presence of organic material and remain active on surface long after application. However they do have a disagreeable odors and cause skin irritation.

Alcohols

Alcohols are most widely used disinfectant and antiseptic, however they are not sporicidal but are bactericidal and fungicidal. The two most commonly used alcohol germicides are ethanol and isoproponol (in 70–80% concentration). They act by denaturation of proteins and dissolution of lipids.

Halogens

A halogen is any of the five elements Fluorine, Chlorine, Bromine, Iodine and Astatine. They exist in diatomic state in free state and form salt like compounds with sodium and other metals. Sodium based formulations are used as skin disinfectant and kills by oxidation of cell constituents. At higher concentration, it destroys spores. Iodine often is applied as tincture of Iodine (2%) on water – ethanol solution of potassium iodine. This form are effective antiseptic but damages skin and results in allergies. Iodine has been complexed with an organic carrier to form Iodophore. Iodophore are water soluble and release iodine slowly to minimize skin preoperative skin degerming or for disinfecting in hospitals.

Chlorine is used for disinfecting municipal water supplies and swimming pools and is employed in dairy and food industries. It may be applied as Chlorine gas, Sodium hypochlorite (HCLO) or Calcium hypochlorite, all of which yield Hypochlorous acid and then atomic oxygen. This results in oxidation of cellular materials and destruction of vegetative bacteria and fungi, but not spores.

$$Cl_2 + H_2O \rightarrow HCl + HClO$$

$$Ca(OCl_2) + 2H_2O \rightarrow Ca(OH)_2 + 2HClO$$

$$HClO \rightarrow HCl + O$$

Death of microbes occurs within 30 minutes. It is in expensive, effective and easy to employ (but chlorine reacts with organic compounds to form toxic trihalomethanes).

Heavy metals

Mercury, Silver, Arsenic, Zinc and Copper were used as germicides (but more recently they have been replaced with less toxic and more effective germicides). 1% silver nitrate is added to eyes of infants to prevent ophthalmic gonorrhea, Silver sulfadiazine is used on burns and Copper sulfate is effective as algicides in lake and swimming pools etc. Heavy metals combine with proteins (sulfhydril group) and inactivate them. They also precipitate cell proteins.

Aldehydes

Aldehydes such as formaldehyde and gluteraldehyde are highly reactive molecules which combine with nitric acids and proteins inactivating them. They are sporicidal and are used as chemical sterilants. 2% buffered Gluteraldehyde is an effective disinfectant in hospitals and laboratories. It readily destroys microbes within about 10 minutes, but requires 12 hrs to destroy spores.

Quaternary Ammonium Compounds

Detergents (Latin deterger = wipe off) are organic molecules which serve as wetting agents and emulsifies because they have both hydrophylic and hydrophobic ends. This amphipathic nature solubilize otherwise insoluble residues.

Although ammonia detergents have antimicrobial property only cationic detergents are effective disinfectants. The most popular of these disinfectants are Quarternary Ammonium compounds characterized by positively charged quartenery nitrogen and long hydrophobic aliphatic chain. They disrupt microbial membranes and denature proteins. Cationic detergents are used as disinfectants for food utensil and small instruments.

Plant pathology is the scientific study of plant diseases caused by plant pathogens the branch of science here deals with the study of plant pathogens its epidemiology symptoms and control is defined as plant pathology organisms that cause infections include fungi, bacteria, viruses, veroids, protozoa etc carriers of these organisms may involve diseased plant parts or insects nematodes common.

Tyndallization

Tyndallization or Fractional sterilization can be accomplished by free flowing steam at a temperature of 100°C. John Tyndall devised the process of sterilization by steaming at 100°C for 20 min for three consecutive days with an incubation period of 24 hrs at room temperature. During the incubation period spores germinate to vegetative cells, which are killed at 100°C on subsequent exposure to heat (If the spores are present and do not germinate during incubation period the method well fail to sterilize the material) An improper sterilization may occur due to media unsuitable for germination of spores and spores of anaerobic bacteria will not germinate if exposed to air: similarly spores of aerobic bacteria will not germinate if not freely exposed to air. Another disadvantage of the method is time consumption.

This method is suitable for sterilization of material which are hydrolyzed or decomposed at a temperature over 100°C. An apparatus known as Arnold sterilizer is used for Tyndallization method

Chapter 9

MYCOTOXINS

MYCOTOXINS

A very large number of molds have been demonstrated to produce toxic substances designated Mycotoxins. Mycotoxins are fungal metabolites. Some are mutagenic and carcinogenic, some display specific organ toxicity and some are toxic by other mechanism. The two predominant genera of fungi in stored products are probably Penicillium and Aspergillus, members of which produce mycotoxins.

The syndrome resulting from indigestion of the toxin in mold contaminated food is referred to as Mycotoxicosis. Historically first documented case of mycotoxicosis attributed to a fungus containing food was that of rye ergot. *Claviceps purpurea* parasitizes rye and other grains and elaborates many lysergic acid derivatives. Consumption of grain or flour made from it over a period of time can result in gangrenous ergotism. Outbreak of ergotism was quite common during middle ages. More recent outbreaks have been reported in Soviet Union (1926) England (1928) France (1951).

AFLATOXINS

In 1960 100,000 turkey poults died suddenly in England. It was followed by reported death of 14,000 ducklings and nine out breaks in calves. A common factor in each of these episodes was use of Brazilian peanut meal as a part of animal feed. *Aspergillus flavus* was isolated from the meal and subsequent toxin was identified. The resulting *A flavus* toxin was called Aflatoxin.

The genus *Aspergillus* is divided into morphologically distinguishable groups, one of which is *A flavus* group containing *A flavus* and *A parasiticus*. The number and types of Aflatoxins produced very with the strain.

***Chemistry*:** AFB1 or Aflatoxin B1 is produced by all aflatoxin positive strains and is the most potent of all.

AFM_1, is hydroxylated product of AFB and appears in milk, urine, feces and other metabolic products.

AFL, AFLH, AFQ_1 and AFP_1 are all derived from AFB_1.

AFB_2 is the 2,3 dehydro form of AFG_1. The toxicity of the six most potent aflatoxins decreases in the following order; $B_1>M_1>G_1>B_2>M_2>G_2$ when viewed under ultraviolet (UV) light, six of the toxins florescence as noted have

B_1 & B_2	-	Blue
G_1	-	Green
G_2	-	Green blue
M_1	-	Blue violet
M_2	-	Violet...

Requirements for Growth and Toxin Production

A flavus and *A parasiticus* grow on a wide variety of substrates over a mesophilic range of temperature. Optimum conditions for the production of Aflatoxin would be an a_w of 0.85 and a temperature of 25–40°C. Aflatoxin synthesis can be inhibited by caffeine, which apparently inhibits the uptake of carbohydrates.

Toxicity

Aflatoxin B_1, the most toxic of aflatoxins is toxic to various animals. Many of the other aflatoxin have been shown to be toxic to different species of fish, mammals and poultry. When cows eat feed containing Aflatoxin, Aflatoxin M_1 and M_2 are excreted in milk. Although Aflatoxin M_1 and M_2 less toxic than parent compound B_1 and B_2, M_1 retain its toxic ability in many animals.

Significance of Toxin in Food

Many commodities will support the growth of toxigenic strains if inoculated, including various dairy products, bakery products, fruit juices, cereals and forage corps, on fresh beef, ham and bacon etc. Peanuts, cotton seeds and corn however differ significantly in that these products are susceptible to fungal infection, growth and mycotoxin production before harvesting. The contamination and potential of aflatoxin production in these crops is related to insect damage, humidity, weather conditions and agricultural practices. In most other cases, the growth of toxigenic strains and elaboration of Aflatoxin occurs following harvest or formulation of product.

Aflatoxin production has been demonstrated to occur on an endless number of food products. Under optimal conditions of growth some toxins can be detected within 24 h (otherwise within 4–10 days). On peanut the following observations have been made.

1. Growth and formation of Aflatoxin ocur mostly during curing of peanuts after removal from soil.
2. In a toxic lot of peanuts, only comparatively few kernels contain toxin and success in detecting toxin depends on collecting a relatively large sample such as 1 kg for assay.
3. The two most important factor affecting Aflatoxin formation are moisture and temperature.
4. The toxin may vary greatly in amount even within a single kernel.

The US Food and Drug Adminstration (FDA) has established allowable levels of Aflatoxins in food as 20 ppb for food, feeds, brazil nuts, peanut etc and 0.5 ppb for milk.

For expression of mutagenicity mammalian metabolizing system are essential for Aflatoxins, especially AFB_1. Also essential is their binding with nucleic acid especially DNA.

While nuclear DNA is normally affected AFB_1 has been shown to bind covalently to liver mitochondrial DNA. Cellular macromolecules other than nucleic acids are possible sites for Aflatoxins. The site of aflatoxin molecule responsible for mutagenicity is C2-C3 double bond in dehydro moiety. Its reduction to 2,3 dihydro (AFB_2) from reduces mutagenicity 200–500 fold. Following binding to DNA, point mutations are predominant genetic lesions induced by Aflatoxins, although frameshift mutations are known to occur.

Most species of susceptible animals die within 3 days after administration of toxins and shows gross liver damage. Aflatoxins leads to EFDV syndrome, Reyes syndrome etc.

SUGGESTED READING

1. Alberts B, Bray D & Johnson A. 1997. Essential Cell Biology. London: Garland Publishing.
2. Allsopp D & Seal J. 1986, Introduction to Biodeterioration, Edward Arnold Pub. London.
3. Atlas RM. 2000. Microbiology Fundamentals and Applications, MacMillan Pub. Co., New York.
4. Casida J.F. 2010. Industrial Microbiology, New Age International India Pvt. Ltd., New Delhi.
5. Chatterji AK. 2005. Introduction to Environmental Biotechnology.
6. Crueger W & Crueger A. 2005. Biotechnology: A Text Book of Industrial Microbiology, II Edn., Panima Publishing corporation, New Delhi.
7. Cruger W, Cruger A & Brock TD. 1991. Biotechnology, A Text book of Industrial Microbiology.
8. Darwin C. 1859. On the Origin of Species. London: Murray.
9. Demain AL & Davies JE. 1999. Manual of Industrial Microbiology & Biotechnology. ASM press.
10. El-Mansi EMT, Bryce CFA, Dahhou D, Sanchez S, Demain AL & Allman AR. 2012. Fermentation Microbiology and Biotechnology. III Edn., CRC Press, London.
11. Flickinger MC & Drew SW. 1999. Encyclopaedia of Bioprocess Technology Fermentation, Biocatalysis and Bioseperation Vol. V., John Wiley and Sons Publications.
12. Glazer AN & Nikaido H. 1995. Microbial Biotechnology – Fundamentals of Applied Microbiology. W.H. Freeman and Company. New York.
13. Graur D & Li W-H. 1999. Fundamentals of Molecular Evolution, 2nd edn. Sunderland, MA: Sinauer Associates.
14. Kalaichelvan PT & Arul Pandi I. 2007. Bioprocess Technology, MJP publishers, Chennai.
15. Madigan MT, Martinko JM & Parker J. 2000. Brock's Biology of Microorganisms, 9th edn. Englewood Cliffs, NJ: Prentice Hall.
16. Markandy DK & Rajvaidys N. 2004. Environmental Biotechnology. APH Publishing Corporation, New Delhi.
17. Mohapatra PK. 2006. Text book of environmental biotechnology, I.K. International publishing house, New Delhi.

18. Norris RD. 1994, Handbook of Bioremediation, Lewis Publishers, London.
19. Patel A.H. 1996. Text Book of Industrial Microbiology, MacMillan India Ltd., New Delhi
20. Peppler H & Pearman D. 2008. Microbial Technology, II Edn. Vol.I, Academic Press, New York.
21. Pragya Khanna. 2008. Cell and Molecular Biology. I.K. International Pvt Ltd, 2008. p. 221. ISBN 81-89866-59-1, 978-81-89866-59-4
22. Prescott LM, Harley JP & Helin DA. 2008. Microbiology, Fifth Edition, McGraw Hill, New Delhi.
23. Rajendran P & Gunasekaran P. 2006. Microbial Bioremediation. MJP Publishers, Chennai.
24. Stanbury PF, Whitaker A & Hall SJ. 1999. Principles of Fermentation technology, II Edn. Aditya Book (P) Ltd., New Delhi.
25. Stilling P. 2009. Ecology – Theories and Applications. IV Edn., Pearson Education Inc. New Jersey.
26. Subba Rao N.S. 2001. Soil microbiology. Raju Primlani Publishing Pvt. Ltd., New Delhi.
27. Waites MJ, Morgan NL, Rockey JS & Higton G. 2001. Industrial Microbiology: An Introduction, Blackwell Science, London.
28. Watson JD, Hopkins NH & Roberts JW et al. 1987. Molecular Biology of the Gene, 4th edn. Menlo Park CA: Benjamin-Cummings.

SUBJECT INDEX

β oxidation 77, 78

A

Abortive transduction 18
Acrocentric 6
Activated sludge process 147, 152, 153, 154
Active site 62, 63, 64, 65
Adsorption chromatography 191, 195
Aerated ponds 156, 157
Aeration 129, 153, 154, 166, 184
Affinity chromatography 196
Aflatoxins 207, 208, 209, 210
Allele 11, 12
Allelic pair 11
Allosteric enzymes 63
Anaerobic digestion 140, 142, 144, 146, 161, 162
Anaerobic ponds 156, 157
Anaphase 6, 9
Antifoams 180, 181, 183
Attenuation 48, 49
Autoradiography 101, 104, 105, 108, 113, 114

B

Baffles 184, 185, 186
Biochemical mutations 50
Biochemical oxygen demand 161
Biogas 162, 164, 165
Biolistics 108, 110
Blotting techniques 102, 103
Buffers 100, 175, 180, 181, 196

C

Catabolite Repression 48
Cell disruption 193, 194
Cesspools 144, 146
Chemical oxygen demand 161
Chlorination 133, 134, 157
Chromatography 102, 191, 195, 196
Chromosome mutation 172
Citric acid cycle 70, 71, 72
Coagulation 193, 124, 129, 130, 147
Coliforms 116, 117, 118, 124
Composite Transposons (Tn) 22
Composting 165, 166, 170
Conditional mutations 50
Conjugation 12, 13, 14, 15, 16
Contact dryers 197
Cosmids 79, 80, 83, 93, 94
Crystallization 178, 192, 197

D

Deletion 50, 52, 114, 172, 189
Diakinesis 8
Dialysis 158, 159
Diatomaceous filters 202
Diplotene 8
Direct Repair 55
Dissolved solids 141, 148
DNA fingerprinting 114

Down stream processing 191
Drying 112, 166, 197

E

Electron microscopy 8, 105, 122, 199
Electrophoresis 89, 98, 99, 100, 103, 104
Electroporation 109
Encystment 123
Enzyme Induction 46, 47
Enzyme Repression 46
Excision Repair 55

F

F' Conjugation 15
F+ × F– mating 14, 15
Facultative ponds 156, 157
Filtration 125, 126, 127, 130, 152, 157
Flocculation 124, 129, 150, 154, 155, 193

G

Generalized Transduction 16, 17
Genetic code 44, 46, 52
Genetic engineering 79, 95
Genome mutation 171
Genotype 11, 12, 13, 50

H

Halogens 204
Head space 180, 184
Helicases 37
Hepatitis 121, 122, 144
Heterozygote 11, 12
Heterogeneous Nuclear RNA 31
Hfr Conjugation 15
Highly repetative sequence 4
Histone 2, 3, 4, 5
Holoenzyme 40, 58
Homozygote 11, 12
HPLC 196
Hybrid vectors 93
Hybridization 102, 104, 111, 112, 113

I

Imhoff tank 144, 145
Induced mutagenesis 53
Inhibitors 60, 61, 182, 196
Insertion 20, 24, 50, 51, 52
Insertion vectors 88
Insertional inactivation 81, 83, 85, 91, 111
Interphase 5, 6, 9
Ion exchange chromatography 196
Ionizing radiations 173, 203, 204

K

Kinetochore 6

L

Lagging Strand 37, 38
Laminar air flow hood 202
Laptotene 7
Leading Strand 37, 38
Leaky Mutations 51
Liposome 111
Liquid shear 193
Lysogenic bacterium 17

M

M 13 90, 91, 92
Maturation ponds 156
Mechanical shearing 97, 98
Meiosis 7, 9, 12
Messenger RNA (MRNA) 29, 30
Metacentric chromosomes 6
Metaphase 6, 8, 9
Michaelis constant 60
Microinjection 108, 110
Mismatch Repair 55
Missense mutations 50, 51
Moderately repetative sequences 4
Molasses 177, 178
Most probable number 126
Multiple cloning site 80, 84, 85
Multiple media filter 131
Multiple Mutations 50
Mutagenesis 50, 53, 173
Mutations 45, 46, 49, 51, 54, 56, 171, 172, 209
Mycotoxins 207

N

Negative supercoiling 2
Non-leaky Mutations 52
Nonsense mutation 51
Northern blotting 102, 104, 106
Nuclear matrix 3
Nucleosome 3, 4

O

Okazaki fragments 35
One gene one enzyme hypothesis 56, 57
Operons 21, 46, 47
Origin of replication 2, 35, 36, 80, 81, 83, 84, 90
Oxidation ponds/lagoons 155
Oxidative phosphorylation 67, 70

P

Pachytene 7, 8
Parasexual cycle 173
pBR 79, 82, 83, 84, 85
Photo Reactivation 55
Plasmids 2, 20, 24, 79, 80, 81, 82, 84
Point Mutations 50, 51, 209
Porcelain filter 202
Positive supercoiling 2
Precursors 104, 134, 175, 182
Pressure filters 131
Prophase 5, 7, 9
Protoplast fusion techniques 174

R

Rapid sand filters 130
Replacement vector 88
Replication forks 35, 37
Reverse osmosis 157, 158, 191, 197
RFLP 113
Ribosomal RNA 2, 4, 29, 47
RNA Primer 35, 36, 37

S

Scaffold Attached Regions (sars) or Matrix Attachment Regions (mars) 3
Screening 85, 91, 98, 99, 112, 128
SEM 201
Septic tank 144, 145, 146
Settleable solids 141, 147, 149, 150
Sex factors 14
Silent mutations 51
Slow sand filtration 131
Solenoid fiber 3
Solid shear 193, 194
SOS Repair 56
Southern blotting 101, 102, 103, 104, 105
Specialized transduction 17, 18
Spontaneous mutagenesis 53
Spray dryers 197
Standard plate count 124
Stock cultures 198
Strain improvement 171
Super helix relaxing protein 34, 36
Supercoiling 2
Suspended solids 126, 128, 132, 138, 141

T

Telocentric 6
Telophase 6, 9
TEM 200, 201
Template DNA 36
Tertiary treatment 148, 157
The mechanism is similar to adsorption 196
Topoisomerases 2
Transcription 4, 21, 38, 39, 40, 41, 47
Transduction 13, 16, 18, 108, 109, 174
Transfer RNA (TRNA) 29, 31
Transformation 13, 14, 16, 19, 24, 66
Transitions 52, 54, 172
Transposons 20, 21, 22, 23, 24
Transversions 52, 172
Trophozoite 123
Tyndallization 206

U

Ultra filtration 191, 196
Ultrasonication 193, 194

Ultraviolet light 173
Unique or non repetative sequence 4
Unique restriction sites 81, 85
Upflow filters 131

W

Western blotting 102, 103, 105
Whey 179
Whole broth processing 198
Wobble Hypothesis 45

Y

YAC 94, 95

Z

Zygotene 7